REFURBISH ANTIQUE TELEPHONES FOR FUN AND HOBBY

Step by step instructions to take an old telephone and return it to its original working order.
No electronics or telephone knowledge needed.

ED MITCHELL

Order this book online at www.trafford.com
or email orders@trafford.com

Most Trafford titles are also available at major online book retailers.

Printed in the United States of America.

ISBN: 978-1-4269-6283-7 (sc)
ISBN: 978-1-4269-6282-0 (hc)
ISBN: 978-1-4269-6284-4 (e)

Library of Congress Control Number: 2011905736

Trafford rev. 05/20/2011

 www.trafford.com

North America & international
toll-free: 1 888 232 4444 (USA & Canada)
phone: 250 383 6864 ♦ fax: 812 355 4082

TABLE OF CONTENTS I

TABLE OF CONTENTS II

NOTE: This table gives the location of the wiring diagrams for all the network and dial types.

CHAPTER 1.
BASIC TOOLS AND TECHNIQUES AND
USEFUL INFORMATION FOR GETTING STARTED

1.1 **Before you begin.** Before getting into the dynamics of refurbishing old telephones, you should be familiar with the basic tools and techniques that are necessary. Nothing expensive is needed; everything is available at local hardware or electronics stores and the more expensive things such as drills and saws are usually available in almost every household anyway as they have general use.

1.2 **Several items used.** Figure 1-1 shows several items used on all phones. In Figure 1-2 are close ups of the spade lugs and crimp-on connectors for the flat telephone cable.

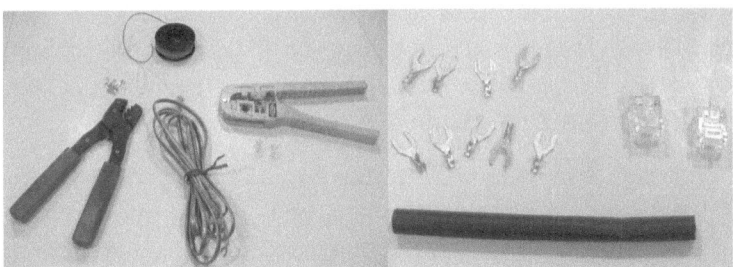

Figure 1-1 (Left). Some items used in refurbishing old telephones. From top going clockwise; 22 gage electronic hook up wire, crimper tool for crimping the connectors onto 4 wire flat telephone cable, flat 4 wire telephone wire, crimpers for crimping the spade lugs to the 22 gage hook up wire.

Figure 1-2 (Right). Close-up of the telephone spade lugs (left), flat cable crimp-on connectors (right), heat shrinkable tubing (bottom). Use 3/16 inch diameter tubing.

1.3 **Hook up wire.** The hook up wire should be #22 gage *stranded* wire, not solid wire, available at Radio Shack or other electronic parts stores. The connectors used for plugging telephones into wall sockets is the four wire crimp on connectors shown in Figure 1-2. The crimping tool shown in Figure 1-1 for crimping the connectors on to the flat cable should be of the same brand. If you cross brands, sometimes they will not be compatible. The telephone 4 wire flat cable can be bought at hardware or electronics parts stores, but if you want it wrapped with brown cotton cloth like with the old antique ones, you will have to order it from one of the suppliers listed in Appendix J. The crimpers for crimping the spade lugs to the hook up wire are available in any hardware store or at Radio Shack. The spade lugs of the size use for telephones may be only available at Radio Shack.

1.4 **Spade lugs.** The spade lugs for the plastic bound hookup wire are designed to work on the #22 gage hook up wire but they do not necessarily work on smaller gage wires. For instance, when you connect a spade lug to one of the 4 wires in the flat cable, the diameter of the wires is much smaller that #22 gage, so you have to strip off a little of the insulation exposing the copper wire underneath, and fold it back over the insulated part of the wire before crimping on the spade connector. See Figure 1-3. The spade lugs used on cotton bound wires used to connect the handsets and earpieces is larger that those used on plastic bound hookup wire and are available from the source referenced in paragraph J.4. Use model A29921CT-ND, 18-20 AWG tin crimp spade lugs.

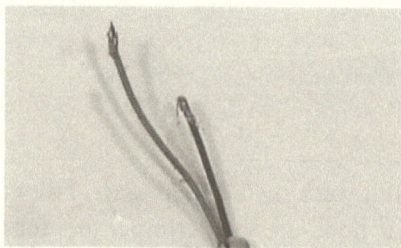

Figure 1-3. Wires from the flat cable prepared for receiving a crimp on spade lug. A little bit of the copper wire must be exposed so it is touching the spade lug after crimping.

Always test your crimp-on connections with a volt ohm meter to be sure you have a good contact. Open wires are probably the hardest thing to find during troubleshooting.

1.5 **Heat shrinkable tubing.** The heat shrinkable tubing is used to insulate a splice after you have soldered the two ends of two wires together. The tubing is designed to shrink in the direction around the diameter but not along the length. When it has been shrunk by applying heat to it with a hair dryer or a heat gun, it makes a good insulator for the splice and is flexible. Use tubing that is 3/16 inch in diameter.

1.6 **More tools.** Figure 1-4 shows some more tools you will need. The #22 gage wire is used to tie down the line cord so it will not pull out of the phone if someone drops the phone or pulls on it too hard. See Figure 7-1.

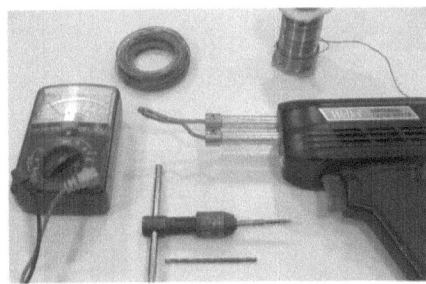

Figure 1-4. Top, left to right: #22 gage wire, solder for electronic circuits; middle: volt ohm meter, soldering gun; bottom center: tap for size 4-40 screws, handle for tap with tap; #43 gauge drill bit.

When purchasing a soldering gun, be sure you get a gun, not a soldering iron. The irons work best on a production line but not for a hobbyist. The volt ohm meter can be an inexpensive analog type. Digital is OK but certainly not necessary. The tap and drill are useful for securing a network or ringer to the bottom of the phone. The bottoms are always steel and you can drill where necessary, then tap the hole to cut threads in it and use a size 4-40 machine screw to secure the network or ringer in place. You will use size 4-40 screws a lot, so buy a bolt cutter as shown in Figure 1-5. The figure shows bolt cutters used to cut any bolt from size 4 through 10.

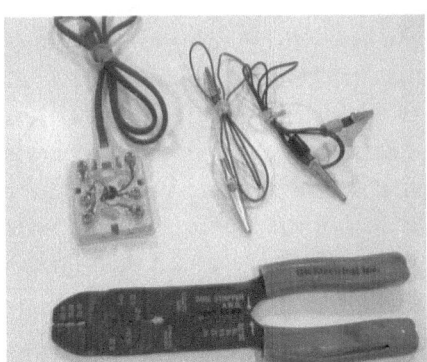

Figure 1-5. Top left; Surface mount modular jack for telephones. Top right; Pair of 15 inch jumper wires useful when testing telephones. Bottom; Bolt cutter for cutting size #4 through #10 size bolts.

It automatically repairs the threads after cutting them to any length you want, so you can get several of the longer ones and cut them to the length you need at the time. The bolt cutter shown is made by GB Electrical, Inc.

1.7 Test wires. At the upper right hand side of Figure 1-5 is a pair of 2 foot long wires with alligator clips on each end. These you just make yourself by buying some alligator clips and hooking them together with 2 feet of wire. These are indispensable when testing circuits and ringers as you can use them to temporarily hook components such as a ringer, or receiver element or transmitter element or anything else to see if it works before you hard-wire it in.

1.8 Surface mount modular jack. Also shown in Figure 1-5 is a surface mount modular jack for telephones that you can buy in most hardware stores or at Radio Shack. Attach a 3 foot flat telephone cable to it, as shown, and you can use it for testing any old telephone that does not have a modern 4-pin jack on it. The red and green wires are attached to screw driver pins so you can easily connect them up to the L1 and L2 terminals on an old telephone, plug it in to a phone line and call that line to see if the phone works. This is a great time saver for testing an old phone for the first time. You don't need to have a complete telephone test platform as described in Appendix A.

1.9 Shock hazard. There is a possible shock hazard while working on telephones that are plugged in to the telephone line, but it is not considered a dangerous shock. There is no shock hazard from the speech part of the electronics; that is, the speaking, receiving, and dialing part. But you can get a shock if someone is calling that line at the same time you are touching the electrical wires where the ring voltage comes in and where it goes to the ringer. The voltage is about 80 volts rms (root mean square) which is less than the voltage you get if you touch the bare wires coming directly out of a household electrical outlet. That voltage is 120 volts rms, and most of us have been shocked by that a few times, and it only resulted in a startling jolt. So the end result is, it will hurt and startle you, but not as much as if you touched an electrical outlet wire. Just make sure you don't have the phone plugged in when you are touching the bare ring circuit wires and you can't get shocked. If you do get shocked, it is not in this author's opinion a big deal. You will learn to avoid it.

1.10 Magnetizer/demagnetizer. When refurbishing old phones you are dealing with many small size screws that are sometimes hard to reach or pick up. It is suggested that you buy a magnetizer/demagnetizer at any hardware store. Follow the directions on the package and you can magnetize or demagnetize your screw driver, which helps you to pick up the screws and place them in hard-to-reach places to screw them in. Be aware that for this to work, the screws must be made of steel. Brass screws will not attach to the screwdriver.

1.11 Will the telephone companies continue to provide rotary dial service? One of the best attractions to having an antique telephone is you can not only have an old antique, but you can use it. There is no predicting when if ever the phone companies will stop providing the rotary dial service, but we do know that the following reasons to continue it are;

- The telephone companies cannot stop it without the permission of their regulating agency, the Federal Communications Commission, (FCC).
- There are still tens of thousands of rotary dial telephone out there that people depend on for their primary telephone use, particularly in the rural areas. The FCC does not want to leave these customers without basic telephone service.

So my guess is that the rotary dial service will be around for a long time.

1.12 Wiring diagram convention used in this book. The wiring diagrams given in chapters 2, 3, 5, 6, 7, and 12 contain the diagrams for the voice electronics only and for reasons of simplicity and clarity, the hookups to the line (L1 and L2) and the hookups for the ringers is omitted as they are given in other chapters. Also, throughout this book, WE means the Western Electric dial and AE means the Automatic Electric dial or its equivalent. Also, the words condenser and capacitor refer to the same thing. Condenser is the old word and capacitor is the newer word.

1.13 Where do you buy old telephones? The usual source is antique stores, the internet, or flea markets. Before the breakup of Bell Systems by the courts in 1986, all individual telephones were owned by the independent privately owned telephone companies, not by the subscriber. When they had an upgrade in technology, the old out-of-date phones were retrieved from the subscriber and destroyed so they would not appear on the black market. Each company had its own method of doing that, but usually they either paid a man to stand there and smash each one with a sledge hammer, or they piled them in a huge pile, poured diesel fuel over them, and burned them. Today, that seems a shame as they are sought after by collectors and command fairly high prices. You can assume with fair accuracy that any old phone dating before 1986 found for sale was stolen from a phone company at some point in time. So be careful when you touch one. It might be hot!

1.14 **Can I short out the switch room?** The answer is no. Luckily, all the outputs from the switch room have high output impedances, which means, the current delivered if you accidentally short one wire to another is very limited. You cannot harm the equipment in the switch room by shorting it, but just make sure you don't accidentally send any high voltages to the switch room.

1.15 **Network designations.** Throughout this book, several different WE networks are referred to and used in wiring diagrams. But you can use any of the following networks; models 425A, 4228, 4227, or 4010B. The differences are so small that they are hardly noticeable to the end user. Also, it is useful to know that the terminals labeled L1, L2, G, F, S, and T, are tie-down points only for conveniently connecting two or three wires together and are not connected to anything inside the network. An exception is the model 101A network, where L1 and L2 are indeed connected to circuitry inside the network.

CHAPTER 2.
OAK WALL PHONE MADE INTO A TOUCH TONE OR ROTARY DIAL TELEPHONE

2.1 **Local battery wall phone.** This chapter will show an old local battery oak wall phone made into a working phone with a modern touch tone dial installed inside, out of view from the outside. The touch tone pad is a Western Electric model 35Y3D and the network is a Western Electric model 425E. Figure 2-1 is a typical phone not refurbished that you might buy in an antique store. Figure 2-2 is the same phone with the door open showing the insides.

Figure 2-1. Old local battery oak wall phone to be converted to a touch tone phone. This one is in good physical shape and does not need refinishing.

Figure 2-2. Insides of wall phone. It contains a magneto, receiver, and hook switch. The compartment where the receiver is laying was the battery box. The owner had to provide his own batteries to make it work in those days, thus the term "local battery" phone. If the phone company provided the power which was the case later when they had newer technology, it was a "common battery" phone.

2.2 **Testing the magneto.** The first thing you do is connect up the magneto to the ringer and see if they work. The magneto is in the middle of the box with the five metal bars and has a crank sticking out the right side. If it has no crank which is quite likely, you must buy one from one of the sources listed in Appendix J. Appendix D shows how to refurbish a magneto and ringer. Also, the original label for the phone should be preserved. That is the paper on the upper door towards the right of the picture. You can preserve it by painting it with some clear sealer such as shellac or wood sealer. If some of it is bubbled up, you must first glue it back down using carpenters glue. It adds a lot of interest and will wear off if not preserved.

2.3 **Testing the hook switch.** After refurbishing the magneto and ringer, make sure the hook switch works. This can be easily done by measuring the two output pins with a volt ohm meter to see if the connections are OPEN when the phone is hung up and CLOSED when it is not hung up. You can easily see how the hook switch works so just look at it and fix what needs fixing by bending the leaf springs with a pair of tweezers. If it cannot be fixed, you must order another from a mail order telephone parts company listed in Appendix J.

2.4 **Installing the dial, network and ringer.** The next step is to install the dial, electronic network, and ringer, if you need it. Make sure that the dial and network work by testing an old 1970's touch tone phone to see if everything works before you take it apart. Always use a Western Electric phone as those are the only ones that this book has the wiring diagrams for. The one shown has a model 35Y3D touch tone (commonly called a TT), and a model 425E network. You can always tell which model touch tone pad you have as it is printed on the sides. They should be taken out of an old 1970's touch tone phone such as the one shown in Figure 2-3 or you can purchase one from the suppliers listed in Appendix J. If the touch tone pad is a model 35Y3D, it will have 8 wires coming out of the back which are: green, black, white, blue, red-green, orange-black, red, and white-blue. If it is a model 72Y3D pad, it will have 9 wires coming out the back with the same color markings as the model 35 but with an additional brown wire. The network may be any WE network, that is, models 425E, 4228, 4227, 424-4T, or 4010B.

Figure 2-3. 1970's touch tone phone. This will be taken apart and the dial and network installed on the old wall phone. Try it and make sure it fully works including the dial before going any further. You don't want to put a phone together with non-working parts.

2.5 **Disassembling the phone.** Turn it over and unscrew the two screws on the back that hold the body on to the base plate. Figure 2-4 shows the phone with the cover off. You should use a Western Electric phone only as that is what this book has the wiring diagram for but many telephones, such as those made by ITT, use parts made by WE so those are OK to use also.

Figure 2-4. Insides of a 1970's touch tone phone. Make certain it works before removing the parts. The transmitter and receiver elements should also be salvaged from the handset as they can be used in other handsets that are to be refurbished.

Figure 2-5. Base of an old phone showing the rivets being drilled out to remove salvageable parts such as the touch tone dial. The two drilled-out rivets are in the direct center of the picture where the two beveled indentations are.

To remove the dial, you must drill out the rivets holding it to the base plate. Figure 2-5 shows this. You might have to use a finishing nail driver to punch out what is remaining of the rivets. Figure 2-6 shows the dial, ringer, network, transmitter element, and receiver element removed from the telephone and handset, and all of these components can be used for refurbishing any old wall telephone.

Figure 2-6. Top row: touch tone dial, receiver element; middle row: ringer, transmitter element; bottom right: electronic network, all salvaged from a 1970's phone.

2.6 **Mounting the dial and network.** The dial and network must now be mounted in the wall telephone. Use the same brackets that come with the dial, but those are mounted on the angle and they must be mounted perpendicular to the dial for this wall phone. To do this, you need a size 4-40 tap and a size #43 drill bit to go along with it, all available at any hardware store. Loosen the screws holding the dial to the mounting brackets, turn them until the brackets are perpendicular to the dial. Then drill a hole with the #43 drill bit and tap it so you can put a size 4-40 short screw from the steel mounting bracket into the side of the dial. When drilling the hole, extreme care must be taken to be sure you don't jam down into the workings of the dial after the drill bit has broken through. If you are using a drill press, this is not a problem, but if you are using just a hand held drill you

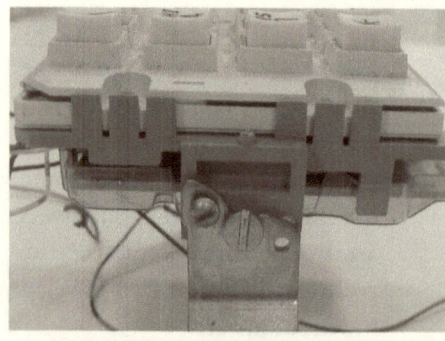

Figure 2-7. Detail of mounting technique to make dial perpendicular to the mounting brackets. A short #4-40 screw is inserted using a drill and a tap to the left of the main screw. This must be done on both mounting brackets.

can easily jam the drill bit into the workings of the dial. So put something like a flat head screwdriver next to the dial to protect this from happening. Tighten all screws down tight as the dial must be solid.

2.7 **Spacer blocks for the dial.** Next the dial and mounting brackets are screwed to the back of the wall phone where the old batteries were located. You can't screw it directly into the back wall, as it would be too far recessed into the box making it hard to reach. You must make spacer blocks of the right thickness to place the keys about ¼ inch below the inside surface of the panel when it is closed. Only then is it comfortable to use. Figure 2-8 shows the mounting method of the touch tone dial on a spacer block of wood to raise

it to just a little below the back of the cover when the cover is closed.

2.8 **Installing the network.** The next part to install is the electronic network. It should be installed close to the dial to minimize the length of the interconnecting wires between them. It is easy to install as it comes with mounting brackets. Figure 2-8 shows it mounted close to the network. You also need to install a two pin terminal strip for connecting wires together and label them S and T. You can use an old terminal strip salvaged from another old phone, or you can buy one from Radio Shack, part #274-688. All you need are two terminals, so the one from Radio Shack can be sawed to the length you want. Figure 2-9 shows a terminal strip salvaged from an old phone.

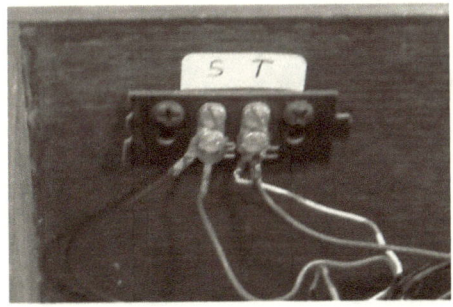

Figure 2-9. Two pin terminal strip salvaged from another old telephone labeled S and T. It should be mounted close to the touch tone pad.

2.9 **Removing the non-wooden parts.** The non-wood parts such as the receiver, transmitter, gooseneck, bells, hook-switch, etc. should be removed from the wooden case and cleaned and painted. Appendix G addresses these tasks.

2.10. **Installing the hook switch.** Now you are ready for the hook switch. Normally the phone box will have one already installed. If not, you will have to buy one and install it. Figure 2-10 shows an installed hook switch.

Figure 2-10. Hook switch mounted on side of old wall phone. If you have to buy a hook switch, it may not look just like the one shown, but if you can fit it in and it has a single pole switch on it, it will work satisfactorily.

2.11 **Wiring the ringer, magneto, and network.** Everything is mounted so the phone is ready to be wired. First, wire up the ringer, magneto, and network and test them before preceding the next step. Figures 2-11 and 2-12 show the wiring diagrams for two and three terminal magnetos. In each case, don't forget to jump L1 to K.

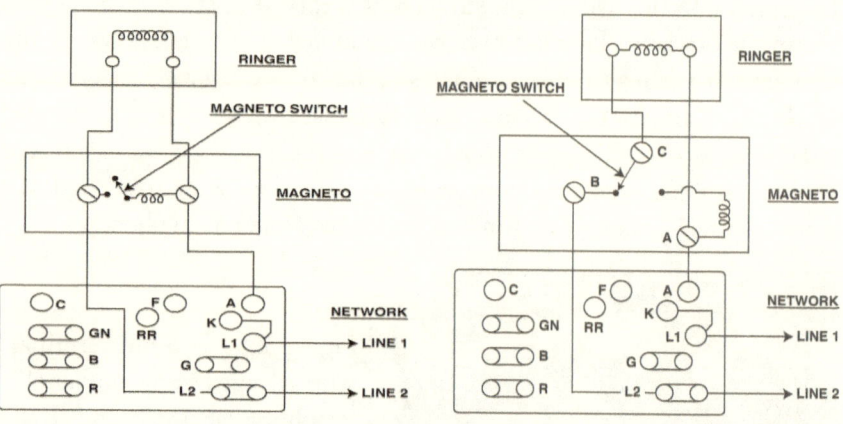

Figures 2-11 and 2-12. Ringer and magneto connection diagrams for two and three terminal magnetos. Two terminals on the left, three terminals on the right.

The one you use depends upon which type of magneto you have. Figures 2-13 and 2-14 show the two types of magnetos.

Figures 2-13 and 2-14. Figure 2-13 has two screw-down terminals at the center, and Figure 2-14 has two at the center and one at the bottom. The black arrows point towards the terminals.

Wire the network, magneto, and ringer up according the wiring diagram in figures 2-11 or 2-12. If you have a two terminal magneto, the wiring is easy, but for a three wire magneto, it is not so easy. To make it easier, take two measurements on the magneto and then label the three terminals A, B, and C. The measurements are:

- While turning the crank, A and C give the magneto voltage (about 60 to 80 volts) as measured using a volt-ohm meter on the AC scale.
- While not turning the crank, B and C are shorted together, as measured using the volt-ohm meter on the ohms scale.

Label the terminals so that both conditions above are met simultaneously and wire up the magneto as shown in Figure 2-12. Then plug the phone line into a telephone jack and call that line using another phone. You don't have to have the rest of the telephone attached to do this test. The ringer should ring when the incoming call is received and it should also ring when you turn the crank. If it doesn't, now is a good time to fix it as you don't have any other wires connected to the network. As a reminder, be sure you jumped L1 to K, as shown on both Figures 2-11 and 2-12. That connection can be easily overlooked.

2.12 **Refurbishing the transmitter and receiver elements.** It is likely that the receiver will work or can be made to work by slightly bending the receiver plate towards the magnets inside the receiver. Refer to Appendix H for instructions for bending the receiver plate until the receiver works satisfactorily. Refer to Appendix E to replace the element altogether with a more modern one if bending the plate does not work. The transmitter element is more of a problem and in most instances will require replacement with a more modern design. Refer to Appendix E for these instructions. The transmitter element has always been the most technically challenging part in the design of a phone until recent times.

2.13 **Wiring the voice part of the phone.** After the ringer is hooked up and working, the rest of the phone is hooked up according to the word descriptions shown below.

For a model 35Y3D touch tone pad:
- CONNECT L2 ON LINE TO L2 ON NETWORK
- CONNECT L1 ON LINE TO L1 ON NETWORK
- ON THE NETWORK, MAKE SURE C IS JUMPED TO L2
- THE HOOK SWITCH GOES BETWEEN L1 AND F
- THE RECEIVER GOES BETWEEN R AND S
- THE TRANSMITTER GOES BETWEEN B AND T
- GREEN TO F
- BLACK TO RR
- WHITE TO GN
- BLUE TO B
- RED-GREEN TO R
- ORANGE-BLACK TO L2
- RED TO T
- WHITE-BLUE TO S
- IF IT HAS A BROWN WIRE, IT IS NOT USED AND MAY BE CUT

For a model 72Y3D touch tone pad:
- DO NOT JUMP C TO L2
- CONNECT L2 ON LINE TO L2 ON NETWORK
- CONNECT L1 ON LINE TO L1 ON NETWORK
- HOOK SWITCH GOES BETWEEN L1 AND F

- RECEIVER GOES BETWEEN R AND S
- TRANSMITTER GOES BETWEEN T AND R
- ORANGE-BLACK TO C
- GREEN TO F
- BLACK TO RR
- WHITE TO GN
- BLUE TO B
- RED-GREEN TO R
- RED TO T
- WHITE-BLUE TO S
- BROWN TO L2. The brown wire is the one that is connected to the tie-down screw on the back of the touch tone pad. It does not go directly to the insides of the pad as the others do.

For a model 25Y3D touch tone pad:
- DO NOT JUMP C TO L2
- CONNECT L2 ON LINE TO L2 ON NETWORK
- CONNECT L1 ON LINE TO L1 ON NETWORK
- HOOK SWITCH GOES BETWEEN L1 AND F
- RECEIVER GOES BETWEEN GN AND S
- TRANSMITTER GOES BETWEEN T AND B
- ORANGE-BLACK TO C
- GREEN TO F
- BLACK TO RR
- WHITE TO S
- BLUE TO B
- RED-GREEN TO R
- RED TO T

Plug the phone in and see if it works. If the dial produces a tone when you touch a key but the phone will not respond and you keep getting the dial tone, this probably means that L1 and L2 have been reversed. This reversal has no effect on a rotary dial phone but it makes a difference in a TT dial. So remove the L1 and L2 wires from the network and reverse them and screw them back on.

2.14 **Polarity of L1 and L2.** On a rotary dial phone, it does not matter which way you have L1 and L2 hooked up. You can have L1 connected to the red line and L2 connected to the green line or vice-versa and the phone will still work. But on a touch tone phone that is not the case. It is easiest if you don't worry about the polarity of L1 and L2 while wiring up the phone, but just reverse the wiring when you test it if you have it wrong. After connecting the phone to the line, if the dial produces a tone when you touch a key but the phone will not respond and you keep getting the dial tone, this probably means that L1 and L2 have been reversed. Just reverse them at the line cable and that should take care of the problem. Figure 2-14 shows a test line polarity reverser that can be easily made. It consists of a coupler and a short section of flat cable with the plug connected the normal way on one end, and the plug turned 180 degrees at the other end. This tester is used until you find out which way L1 and L2 should be wired. Then remove the tester and make the wiring permanent.

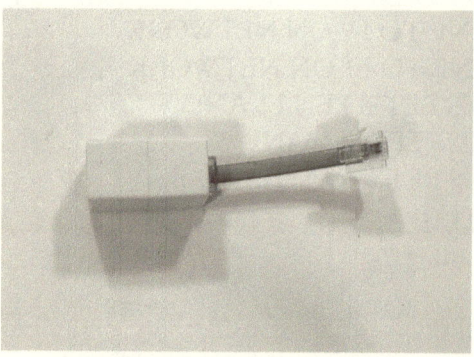

Figure 2.-14. Easily made polarity reverser that can be inserted in the line to reverse the polarity.

2.15 **Troubleshooting.** If it does not work, you must start troubleshooting, and the first thing to do is re-check your wiring. If you are using a model 35 touch-tone pad and a model 4228 network, the wiring is the same as given above but you don't have to mount a 2 position terminal strip for the S and T terminals as the network has those mounted on it already. Testing the TT dial may be done by referring to paragraph A.9.

Any WE network can be used with the model 35 touch tone pad except the model 101 network, but if the network does not have the S and T terminals, simply add them with a 2 position terminal strip shown in Figure 2.9. Remember, the S and T terminals are simply tie-down points and are not connected to anything inside the network.

2.16 Using a rotary dial. If you are using a rotary dial, everything is the same except you don't need a two position terminal board as described in paragraph 2.8, and use the wiring diagrams shown in Figures 5-4, 5-5, or 5-6, depending on whether you are using a WE dial, an AE dial, or no dial.

2.17 Door latch. Since the door must be opened and closed every time you use the phone, you should install a door latch to keep it closed when not in use. The best one is a magnetic latch that you can buy at any hardware store. Get a brown one so it is not too noticeable. The door can be opened by just pulling on it.

CHAPTER 3.
WIRING A DESK SET PHONE USING A WE MODEL 101A NETWORK AND A WE OR AN AE ROTARY DIAL

3.1 **Overview.** This chapter will describe refurbishing an old phone with a model 101A network and either a WE dial or an AE dial. This will normally be a model 102, a model 202, a space saver, or a candlestick phone. There are lots of these around and they are usually in very poor non-working condition with parts missing, no ringer box if it was originally built with one, and frayed cords.

3.2 **A 101A network and an AE dial**. Figure 3-1 shows a model 202 phone with its accompanying ringer box. Every model 202 phone had a ringer box as the technology

Figure 3-1. A model 202 phone with accompanying ringer box before refurbishing.

had not advanced yet to putting all the components in a single desk set. The ringer box has to be refinished and that should be done according to the instructions in Appendix F. The non-wood parts must be cleaned and re-painted according to the instructions in Appendix G.

3.3 **Mounting the network.** Figures 3-2 and 3-3 show the model 202 phone and

Figures 3-2 and 3-3. Insides of the model 202 phone and the ringer box after refurbishing.

ringer box insides after they have been refurbished. Note on the ringer box of Figure 3-3, the model 101A network was mounted on the inside of the lid. This was the only place where there was enough room to fit it in. Figure 3-4 shows a close-up of the model 101A

Figure 3-4. Model 101A network mounted on inside lid of the ringer box. Note the two capacitors in parallel connected to "C" on the network. Each capacitor is 1 microfarad so connecting the two in parallel makes it the 2 microfarad capacitor as specified on the schematic diagram.

network after it had been mounted inside lid of the ringer box.

3.4 **Wiring and testing the ringer**. First wire up just the ringer so you can test it and make sure it works without all the other wiring. The wiring diagram for it is shown in Figure 3-5. The capacitor should be a 0.46 microfarad capacitor or larger. To test the ringer at this stage, plug it in to a phone line and call that phone line using another phone. The ringer should ring. If it doesn't, now is the time to check out the wiring to see if all the wires are conducting and it is wired correctly.

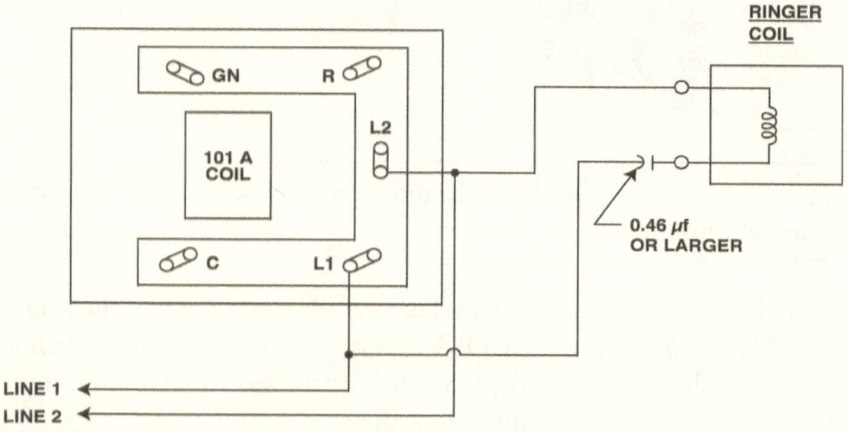

Figure 3-5. Wiring diagram for connecting a 101A network to a ringer. If the phone you are working on has a network instead of a 101A coil, see Figure 4-3.

3.5 **Wiring a 101A network and an AE dial.** Figure 3-6 shows the wiring diagram for a 101A network and an AE dial.

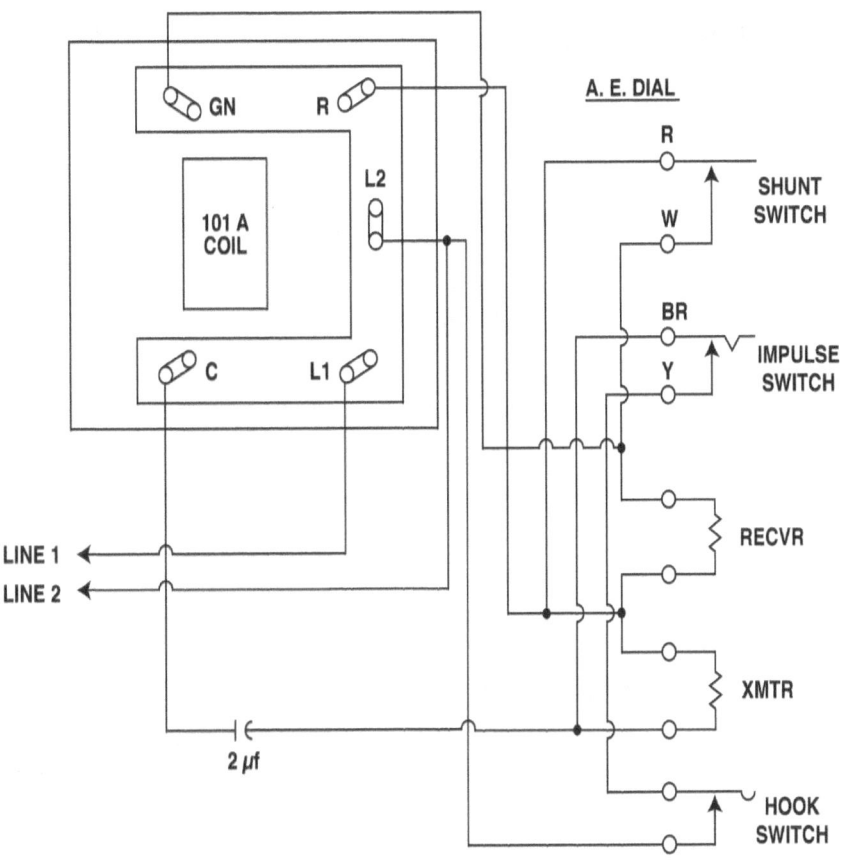

Figure 3-6. Wiring diagram for connecting a 101A network to an AE dial.

3.6 **Wiring a 101A network and a WE dial.** The wiring diagram for the 101A and a WE dial is shown in Figure 3-7. Note that the connection from the transmitter bypass switch to BK is shown as a dashed line because it is made internally. You will not see it on the outside of the dial.

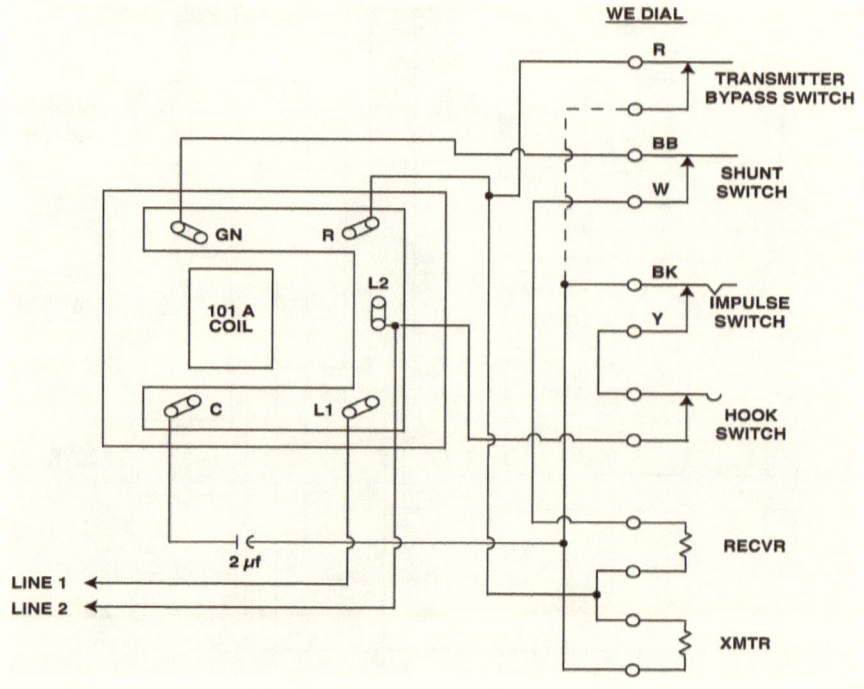

Figure 3-7. Wiring diagram for a 101A network and a WE dial.

3.7 **Wiring a 101A network and no dial.** Figure 3-8 shows the wiring diagram for a 101A network and no dial.

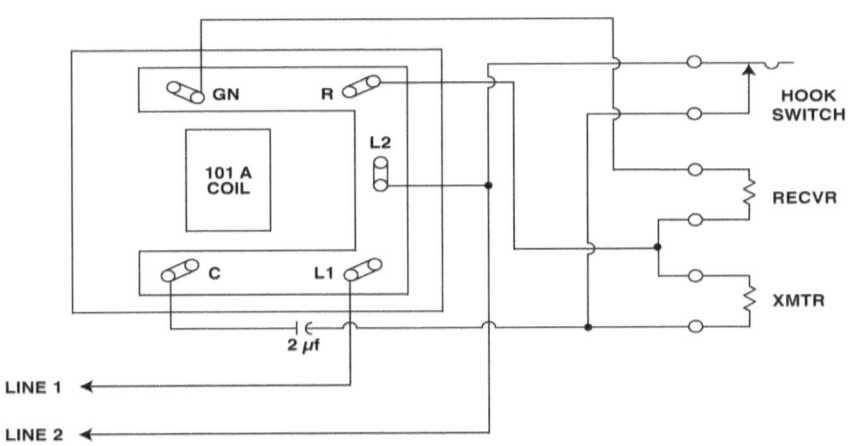

Figure 3-8. Wiring diagram for a 101A network and no dial.

3.8 **The "Wiring Diagram" versus the "Word Description" methods of wiring up a phone.** There are two methods of describing how to wire up a phone, or any other kind of electronic device. The "Wiring Diagram" method as shown in Figures 3-6, 3-7, and 3-8 or the "Word Description" method. The Word Description methods are given here and are the equivalents of the "Wiring Diagram" methods given above. Each reader can choose for himself which method he prefers, but if you have never done it before, I suggest the "word description" method. Some may find one method easier than the other. It is simpler to follow if you first configure the handset from a 4-wire configuration to a 3-wire configuration as is shown in Figure 5-7 and described in Chapter 5.

3.9 **The "Word Description" method of wiring the three circuits above.**

WIRING A 101A NETWORK TO AN AE DIAL (equivalent to the wiring diagram in Figure 3-6 above). R and W are the shunt connections and Y and BR are the impulse connections.

- L1 on network to L1 on incoming line
- L2 on network to L2 on incoming line and one end of hook switch

- Other end of hook switch to Y on dial
- Common on handset to R on dial and R on network
- GN on network to receiver and W on dial
- Transmitter to one end of 2 microfarad capacitor and BR on dial.
- Other end of 2 microfarad capacitor to C on network

WIRING A 101A NETWORK TO A WE DIAL (equivalent to the wiring diagram in Figure 3-7 above). W and BB are the shunt connections and Y and BK are the impulse connections.

- L1 on network to L1 on incoming line
- L2 on network to L2 on incoming line and one end of hook switch
- Other end of hook switch to Y on dial
- Common on handset to R on network and R on dial
- GN on network to BB on dial
- Transmitter to one end of a 2 microfarad capacitor and BK on dial
- Other end of 2 microfarad capacitor to C on network
- Receiver to W on dial

WIRING A 101A NETWORK AND NO DIAL (equivalent to the wiring diagram of Figure 3-8 above).

- L1 on network to L1 on incoming line
- L2 on network to L2 on incoming line and to one end of hook switch
- Other end of hook switch to a 2 microfarad capacitor and one end of transmitter
- Other end of 2 microfarad capacitor to C on network
- Common on handset to R on network
- GN on network to other end of receiver

3.10 Wiring diagram for a 101A network, WE dial and 2 hook switches. Figure 3-8 is a wiring diagram for a 101A network, a WE dial and 2 hook switches. You probably will not be building this combination as it is easier to just use one hook switch, and they work just as well. It is given here so you can use it to repair an existing WE telephone that has 2 hook switches. Note that in Figure 3-7, the connection from the transmitter bypass switch to BK is shown as a dashed line as it is made internally. You will not see it on the outside of the dial.

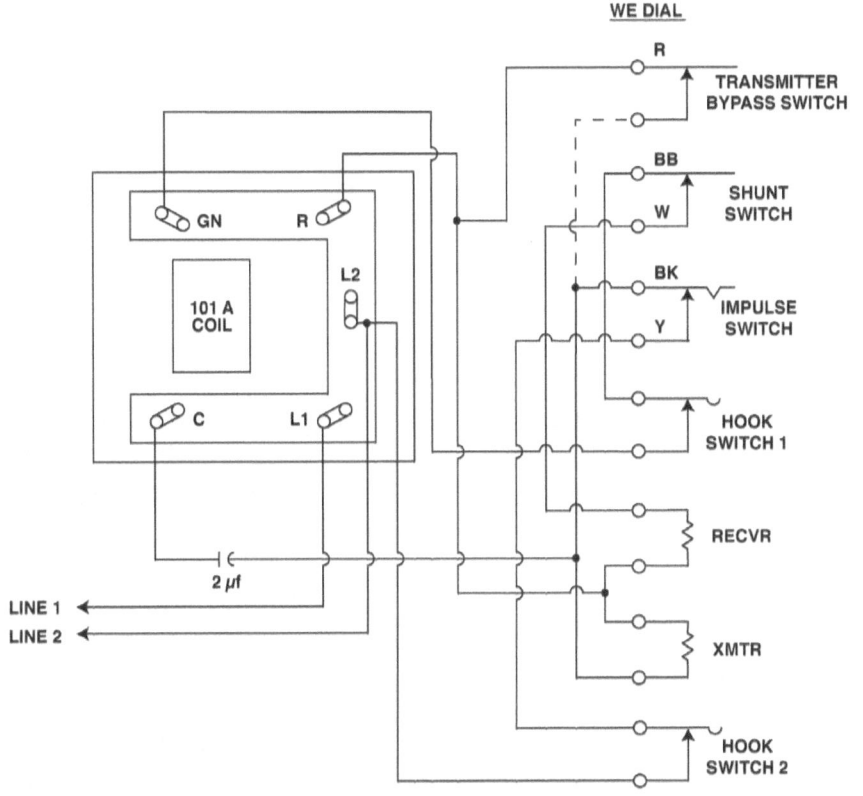

Figure 3-8. Wiring diagram for a 101A network, WE dial, and 2 hook switches.

3.11 The two capacitor WE can. In every model 102, 202, and 302 WE phone there is a long narrow can containing two capacitors; one is the ringer capacitor, and the other the DC blocking capacitor in the receiver circuit. These are shown in Figure 3-10 and the color coding of the wires are shown. Since this capacitor can dates back to as much as eighty years ago, sometimes the colors on the wires are difficult to distinguish. Use your best judgment and if you can get three out of the four, the other can be inferred.

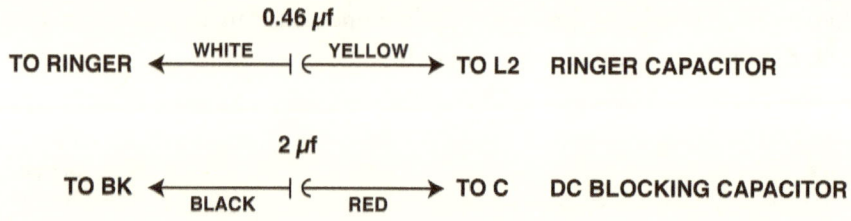

TO RINGER ←——WHITE——┤(←——YELLOW——→ TO L2   **RINGER CAPACITOR**

0.46 µf

2 µf

TO BK ←——BLACK——┤(←——RED——→ TO C   **DC BLOCKING CAPACITOR**

Figure 3-9. Wiring diagram for two capacitors in models 102, 202, and 302 WE phones with color coding of the wires.

3.12 **Unscrewing stuck transmitter and receiver caps.** Many times the transmitter and receiver caps will be stuck as they have not been removed for many years. The best way to unscrew them is not by brute force, but by using heat. Try not to use a heat gun as the heat is applied unevenly, but use water as shown in Figure 3-10. Heat water to just under the boiling point and dip the cap into it as shown and let it soak about 20 seconds. Then take it out, place a rag around it so you won't burn your hand, and unscrew it. If the cap slips inside the rag, cut a strip of fine sandpaper, about 220 grit, into a strip about 5 inches long and 1 inch wide. Place it grit side against the cap, then the rag, and unscrew it.

Figure 3-10. Heating a cap using hot water to allow it to be unscrewed easily (hopefully).

3.13 **Adding a dial to a WE model 302 telephone that has no dial.**
It is common to find a WE model 302 phone with a 101A network but
no dial in it. These always had a round blank panel where the dial would
normally fit with wires attached to a terminal strip located on the back
of the blank panel so it could later be converted to a dial phone by a field
repairman right in the subscriber's home. To install the dial in place of the
blank panel, remove the blank panel, detach the wires connected to the
terminal strip mounted on the back, and install the WE dial. The wiring
diagram to connect the wires to the new dial are included here. Note that
this is only for a WE dial. It won't work for an AE dial.

- Y TO BROWN-YELLOW
- BK TO GRAY AND TO BLACK
- W TO WHITE
- BB TO BLUE-GREEN
- R TO RED-GREEN

CHAPTER 4.
REPLACING THE LINE CABLE AND RE-WIRING THE TELEPHONES FROM THE 1930'S, 1940'S AND 1950'S THAT WERE USED ON PARTY LINES

4.1 **Replacing the line cable.** All of today's telephone companies use a standard wall connector that is not compatible with the plugs from any of the old telephones. So you have to convert them over by installing a flat four wire cable that you can get at any electronics parts store. Be sure to get either a black one from an electronics store, or a brown cloth covered one that can be purchased from the parts suppliers listed in Appendix J. Figure 4-1 shows the old and new flat cable. Always connect the green to green and red to red. Throughout the rest of this chapter, the wiring diagrams do not show L1 and L2 on the network connected to the line cord L1 and L2 going to the phone jack, but of course, you have to connect them to make the phone work. They are left out for the sake of making the wiring diagram less cluttered.

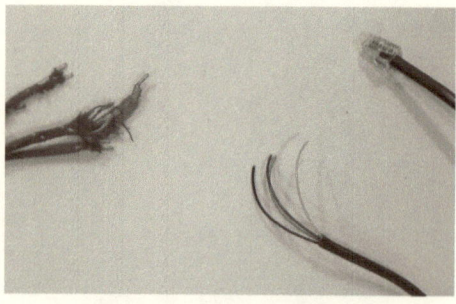

Figure 4-1. The old round cable, left, and new four wire flat cable, right, that has the plug compatible with today's telephone wall jacks.

4.2 **Party lines of past decades.** During the decades of the 1930's, 40's and 50's and 60's the telephone companies used what they called "party lines" to expand their customer base without adding new lines. They would string up to 8 customers on one line and make them time share the line. This, of course, meant that each customer's ring would ring in each of the other 7 household also, which was more aggravating than having to share the voice line. So they devised ways to cut that down to 4 instead of 8 by making one of two design changes. These changes were:

- Install 4 telephones that had ringers that would only ring on one frequency and 4 that would only ring on another frequency, and then select only one frequency at a time; or,

- Wire the phones so that the switch room could control which phones would ring. Four would ring together and four different ones would ring together.

Thus, a customer would only have to listen to four different ringers instead of eight.

If you get an old phone that has not been used since those decades and you want to refurbish it, you must change it back to its original condition so it will ring on every incoming call. If your phone has the re-wiring method, that is, the 2nd method above, you can re-wire it back to make it work. If it has a ringer that uses a different frequency, the 1st option above, the ringer cannot be salvaged and you must install a new one. Figure 4-4 on the bottom left shows a ringer out of a WE princess phone and it is a good choice for replacing the old one in that it is fairly small and has a pleasant sound. These can be purchased from the parts suppliers listed in paragraphs J.2 and J.3 in Appendix J.

4.3 **Phones needing changing.** Figure 4-2 shows a WE model 302 phone from the 30's and '40's, and a WE model 500 phone from the 50's and 60's that were used during the party line era and may have to be changed. These are just the WE models and other manufacturers' phones that are similar in time to these may also need changing.

Figure 4-2. Left: WE model 302 phone. Right: WE model 500 phone. Each may need changing to ring on every incoming call.

To do this you must first plug it in to see if both the voice and ringer circuits work. If the ringer does not work you almost certainly have one out of a 'party line' that needs changing. It is possible that the ringer just does not work but that is not likely as the phones are only about 80 years old at most and they were built very ruggedly. So if the ringer does not work, disconnect the ringer only wires and wire them up as shown in Figure 3-5, if it has a model 101A network, or Figure 4-3b if it has a model 425A network, making sure all the other wires which are associated with the voice circuit are left the same.

4.4 **Re-wiring the ringers.** The four-wire ringer is installed only in the WE model 500 phones and these were only made to work at one frequency, so if you re-wire it according to Figure 4-3 it should work correctly. Figure 4-4 shows typical 2-wire and 4-wire ringers that were made by WE and other brand ringers are similar. If the phone you are working on has a WE model 101A network, use the wiring diagram show in Figure 3-5, in Chapter 3.

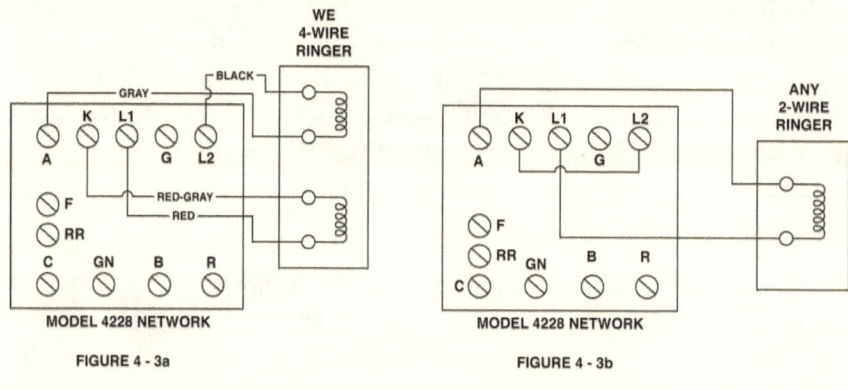

FIGURE 4 - 3a

FIGURE 4 - 3b

Figure 4-3. Wiring diagrams for a four-wire ringer, (figure 4-3a), and a two-wire ringer (figure 4-3b), and a 4228 network.

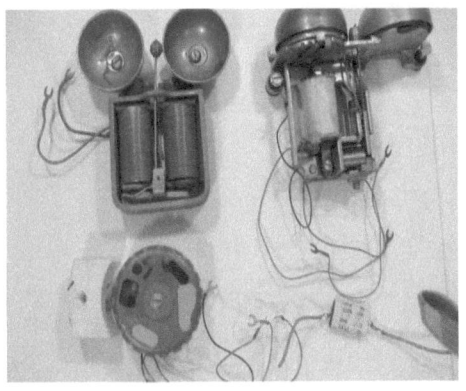

Figure 4-4. Top left: A two-wire ringer from a WE model 302 phone which may have a changed frequency. Top right: a WE 4 wire ringer from a WE model 500 phone. Bottom left: a ringer from a WE princess phone which can be used to replace a ringer that does not work. Bottom right: a typical ringer capacitor.

Two-wire ringers could have been altered by changing the frequency at which they will ring. If this is the case, they cannot be salvaged and must be replaced. A good replacement ringer is the one shown in the lower left corner of Figure 4-4. This round ringer is out of a WE princess phone but can also be ordered from the sources listed in Appendix J. If you use this ringer, it is most important that you also include, if at all possible, the resonator box shown to the left of it in Figure 4-3. The resonator box is not electrically connected to the ringer but it has an *enormous* effect on the perceived loudness, pleasantness of the sound, and length of the after-ring of the ringer. So include it if at all possible and if you have the space! At the bottom right is a typical ringer capacitor of 0.46 microfarads or more. Use it if your phone has a WE model 101 A network that does not have an included ringer capacitor, and wire it up as shown in Figure 8-9 in Chapter 8, except use the ringer in place of the warbler.

CHAPTER 5.
USING TELEPHONE NETWORKS TO
REFURBISH OLD TELEPHONES

5.1 **Different types of networks.** There are seven items you need to make a telephone; a transmitter, receiver, ringer, hook switch, condenser, network, and dial, but the dial is optional. So as long as you keep these 6 or 7 items in a housing, you can make a working phone. Many of the old phones prior to the 1930's utilize the old coils, transmitters and receivers that are no longer useable and must be replaced. So to refurbish these you must replace the coil with a network out of the 1950's; the transmitter element and receiver element with ones from a later era, and the condensers with new ones or condensers that are built into the network. Figure 5-1 shows 7 different networks commonly used for this purpose.

Figure 5-1. Seven commonly used networks from telephones. Top row from left; WE model 101A, WE model 425E, network/warbler for candlestick phones. Center; Stromberg Carlson network. Bottom; WE model 4228, WE model 4227, mininetwork.

This chapter shows how to use the WE models 425A, 4228, 4227, and 4010B networks to replace the existing electronics in any phone, either American made or foreign. Any of these models may be used using the wiring diagrams given in this chapter because electrically they are approximately the same. If there are any differences, they are hardly noticeable to the end user.

5.2 **Useful to know characteristics of WE networks.** Anyone working with telephone networks should know that not all the screw-down posts on the network are connected to the circuitry inside the network. In fact, six of them were put there just as conveniently located tie-down points where external wires could be connected together but not to anything in the circuit. There are: L1, L2, G, F, S, and T. This does not include the 101A network. The L1 and L2 posts on the 101A network are indeed connected to circuitry inside.

5.3 **Determining whether the phone already works.** First, you should determine that the phone does not work with its existing electronics. To do that, connect the phone to a telephone line and listen for a dial tone. First, you will have to replace the line to the phone jack as discussed in paragraph 4.1. If it is a US phone, L1 and L2 are always marked but if it is a foreign made phone, this is not always the case. If yours is a foreign phone where the line wires are not marked, somebody had to disconnect them, so sometimes the screws are loose and all the others are tight, so those must be L1 and L2. If not, and there are no other signs, then you have no choice but to replace all the electronics.

5.4 **Mounting the network.** There are two reasons that will require replacing the existing electronics with a more modern network:

- When you don't have a wiring diagram for the phone.
- When it won't work because it has the old coil, which has lost its' magnetism over time. This results in a loss of transmitter and receiver volume.

Figures 5-2 show an Ericsson phone that works fine even with the old coil so no modification is needed. Figure 5-3 shows a similar phone with the coil removed and a network mounted in its place.

Figures 5-2 and 5-3. Two similar Ericsson phones. In figure 5-2 the old electronics with the original coil is left in place as it works fine. In figure 5-3 the old coil no longer works so it was replaced with a WE model 4228 network.

After you have all the essential elements physically located in the phone, wire up the L1 and L2 line cables from the wall connector and wire up the ringer and condenser, and then plug it in to a telephone line and call that line to make sure the ringer works.

5.5 **Wiring up the phone.** The wiring diagram for connecting up the network to the existing other components is shown in figures 5-4, 5-5, and 5-6 depending on whether you have a WE, AE, or no dial. Don't forget to jump L1 to RR. Not included in the diagrams is the connection of L1 and L2 to the telephone line L1 and L2. They are left out to simplify the diagrams but of course they have to be included. Also, remember that the model 4228 network can also be any of the networks listed in paragraph 1.15.

A WE dial has an "R" connector that is used to short out the transmitter when dialing so the dial pulses don't have to go through the transmitter element. This is a nice

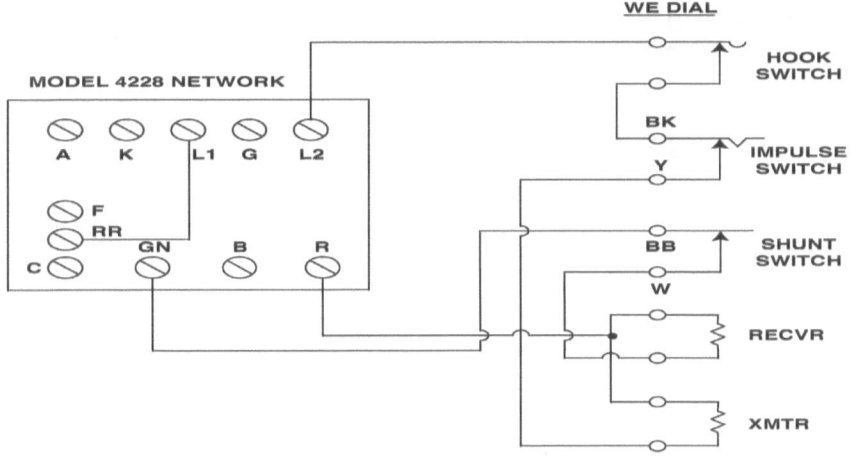

Figure 5-4. Wiring diagram for a 4228 network and a WE dial. Don't forget to jump L1 to RR.

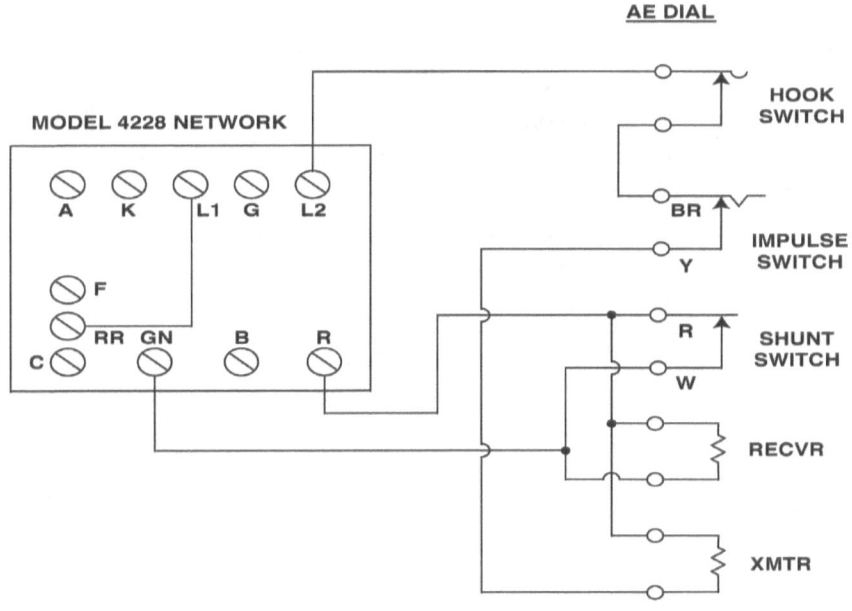

Figure 5-5. Wiring diagram for a 4228 network and an AE dial.

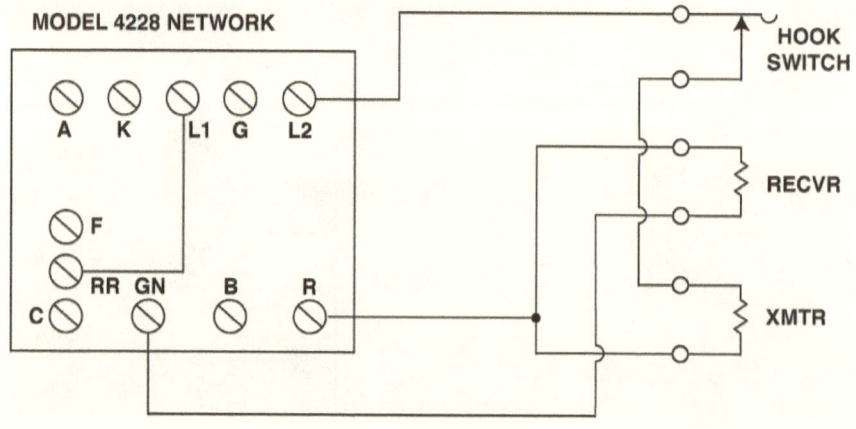

Figure 5-6. Wiring diagram for a 4228 network and no dial

feature to have, but the phone works just as well if the R connector is not connected up. The WE company itself even abandoned the idea when it came out with the "Princess" telephone. So it is left out here to simplify the wiring and make it easier to connect the network to the dial when they are physically separated from each other, such as in any phone with a ringer box that is remote from the desk set. If you use the "R" feature it takes five wires instead of three to connect the ringer box and the desk set. The AE and similar manufacturers' dials never had that feature. Also, the diagrams are given for a WE model 4228 network but you can use any WE network that has the L1, L2, F, A, K, RR, GN, C, and R connectors on it, such as the models 425A, 4228, 4227, or 4010B.

5.6 **Terminal labels on the WE and AE dials.** Every WE dial has the connectors labeled with BK, Y, BB, etc. but the AE dials do not. That is not a problem as there are only 4 connectors on an AE dial that you need: two for the impulse switch, and two for the shunt switch. You can tell which one is which by looking at them; the impulse switch is the one that continually opens and closes as the finger wheel is winding down and the shunt switch is the one that closes at the start of the motion of the finger wheel, and opens at the end of the motion of the finger wheel . On the WE dials, sometimes you can't see the switches so you go by the labeling. The AE dial actually has 5 or more connectors as you can see by referring to Figures B-1 and B-2, but you only use 4 of them. To tell which ones, the two connected to the impulse switch are Y and BR, and the two that short together when the dial is winding up and down are R and W. The others are not used. These can be distinguished by using your volt ohm meter.

5.7 **The "Wiring Diagram" versus the "Word Description" methods of wiring up a phone.** There are two methods of describing how to wire up a phone, or any other kind of electronic device. The "Wiring Diagram" method as shown in Figures 5-4, 5-5, and 5-6, or the "Word Description" method. The Word Description methods are given here and are the equivalents of the "Wiring Diagram" methods given above. Every reader can choose for themselves which method the prefer. Some may find one method easier that the other. It is simpler to follow if you configure the handset from a 4-wire configuration to a 3-wire configuration as is shown in Figure 5-7.

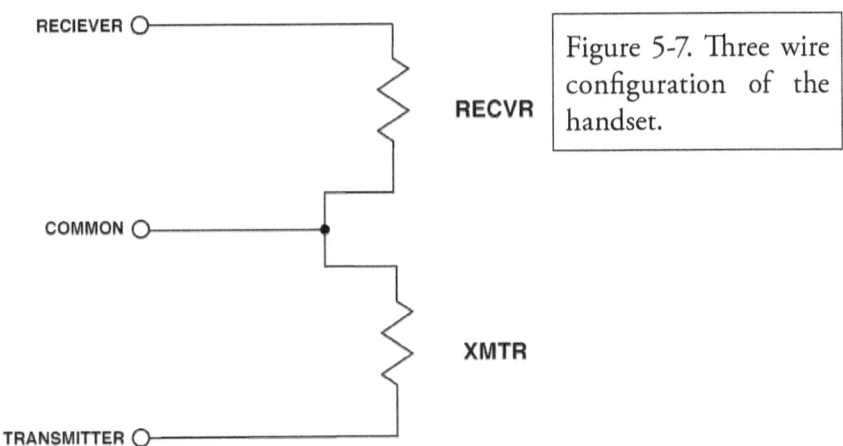

Figure 5-7. Three wire configuration of the handset.

You just tie one end of the receiver and one end of the transmitter together in the handset and call it a COMMON connection. This can be done either in the handset and have only 3 wires connecting to the desk set, or bring all 4 wires to the desk set and tie two of them together there. Either way, the handset is ready to be wired to the other components.

5.8 **The "Word Description" methods of wiring the three circuits above.**

WIRING THE 4228 NETWORK USING A WE DIAL (equivalent to the wiring diagram in Figure 5-4).

- Jump L1 to RR, both on the network
- L1 on network to L1 on incoming line

- L2 on incoming line to L2 on network and to one end of hook switch
- Other end of hook switch to BK on dial
- Common on handset to R on network
- W on dial to receiver
- GN on network to BB on dial
- Y on dial to transmitter
- There is no connection to R on dial

WIRING THE 4228 NETWORK USING AN AE DIAL (equivalent to the wiring diagram in Figure 5-5).

- Jump L1 to RR, both on network
- L1 on incoming line to L1 on network
- L2 on incoming line to L2 on network and to one end of hook switch
- Other end of hook switch to BR on dial
- Common on handset to R on dial
- R on network to R on dial
- GN on network to W on dial and to receiver
- Y on dial to transmitter

WIRING THE 4228 NETWORK USING NO DIAL (equivalent to the wiring diagram in Figure 5-6).

- Jump L1 to RR, both on network
- L1 on incoming line to L1 on network
- L2 on incoming line to L2 on network and to one end of hook switch
- Other end of hook switch to transmitter
- Common on handset to R on network
- GN on network to receiver

5.9   The "Repeating R" factor. Everyone that works with old telephones should know about the "Repeating R" factor. There are three places where the letter "R" is used to represent an electrical point and they are electrically <u>not</u> the same point. The three places are:

- Half of the shunt switch of an AE dial. W is the other half
- The connector on a WE dial used to short the transmitter out during dialing so the dial pulses don't go through the transmitter on their way to the switch room.
- The connector on a WE network that ties the common lead of the transmitter and receiver to the networks' internal electronics.

When you read about the R connector, you have to know whether the author means the one on the network, the one on the WE dial, or the one on the AE dial. Usually you can tell by the context of the rest of the message. If the AE dial is being discussed, then it means the one on that dial, etc. If an author is talking about the R connection and he is not referring to any particular part of the phone, he should be courteous enough to tell you which one he is talking about, but don't bet on it. You might have to figure it out for yourself.

5.10 Three wire versus four wire connections to a remote ringer box. Note on Figures 5-4, 5-5, and 5-6 that there are only three wires needed between the network which is in one location such as in the ringer box, and the dial, hook switch, transmitter, and receiver which is at some other remote location such as in the desk set. Thus, you only need a three wire cable to connect the two together, and most antique cables are three or four wires. So the 4228, 4227, 4010, and 425 networks are good choices if you are wiring up an old phone with a remote ringer box. The WE model 101A network takes 4 wires, so it is OK to use a 101A network as long as you have a 4 wire cable to connect the ringer box and desk set, but you cannot use a 3 wire cable.

CHAPTER 6.
REFURBISHING CANDLESTICK TELEPHONES

6.1 **Using a network/warbler combination, a network, or a modified mini-network.** There are three ways to refurbish candlestick phones;

- Using a network, as described in paragraphs 6.1, through 6.4.
- Using a network/warbler designed specifically for candlestick phones, as described in paragraphs 6.5. and 6.6.
- Using a modified mini-network, as described in paragraph 6.7.

6.2 **Using a network.** Figure 6-1 shows an antique candlestick with an original ringer box.

Figure 6-1. Antique candlestick telephone with original ringer box. They are connected together by a 5 foot cable. The ringer box is mounted on the wall down by the telephone connector and the desk set is up on the desk or table.

The name "ringer box" is a misnomer as it actually contains *most* of the electronics for the phone, not just the ringer. It houses the ringer, condenser, and network, while the desk set houses the transmitter, receiver, hook switch, and dial. They are separated by a 5 foot cable. All telephones had so-called ringer boxes up until about the time the WE model 302 phone came out in the 1940's. With that phone, the technology had advanced to the point where they could make all the components small enough to fit into the desk set. That was a really novel idea at the time and considered a significant advance in technology. But the candlesticks were in use before that so they all had a ringer box. Figure 6-2 shows the insides of the ringer box. The original ringer, condenser, and coil have been removed and replaced by a ringer and model 4010B network, both out of a WE model 500 or similar phone or purchased from one of the suppliers listed in Appendix J.

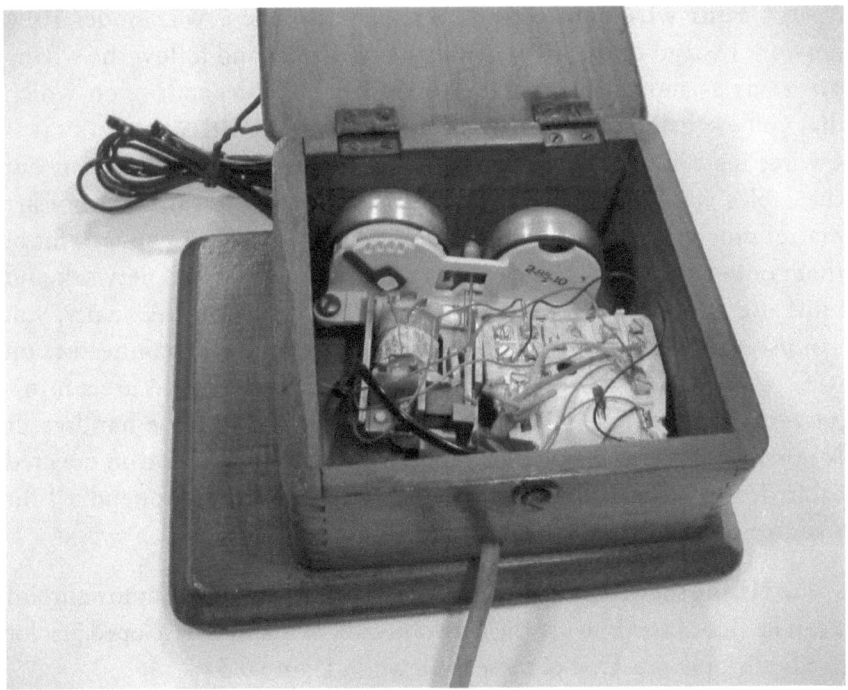

Figure 6-2. Insides of a candlestick phone ringer box. The ringer, condenser, and coil have been removed and replaced with a ringer from a WE model 500 telephone and a network from a WE "Princess" phone. The condenser is imbedded inside the network.

These are wired up as shown in either Figure 5-4, 5-5, or 5-6, depending on whether you are using a WE, AE, or no dial. It is interesting to know that when people saved old candlestick phones, they usually did not save the ringer box as it wasn't considered important. It is important indeed as it has most of the electronics, so it is much harder to find the old ringer boxes than the desk sets.

6.3 **Three wire connection.** Note in the above referenced figures, there is only a need for three wires going from the network, which in this case is in the ringer box, to the rest of the phones' components, which is in the candlestick phone desk set. This means you need a cable connecting the two that only has three wires in it. These are easily available and you can use either a black cable or a brown cotton bound cable available from the suppliers listed in Appendix J.

6.4 **Four wire connection.** You can also use a WE model 101A network instead of the 4010 or similar networks and follow the wiring diagrams as shown in Figures 3-6, 3-7, or 3-8, depending on which dial you are using or if you are using no dial. The 101A network takes 4 wires instead of three to connect the ringer box to the desk set, but the cables you use usually have 4 wires in them anyway, so it doesn't matter much. If you are using a black cable from an old phone, a lot of them only have 3 wires, so you can't use the model 101A network and must use a model 425E, 4010, 4227, or 4228 network. Actually, you can use any WE network that has at least the following connectors on it; C, GN, R, RR, F, A, K, C, L1, and L2. If you want to use a mini-network, it takes 7 wires between the ringer box and the handset. It is possible to do it as you can purchase 8 wire brown cotton covered cable. It is a little bit big but it is a possibility. Refer to Appendix J for a source of this cable.

6.5 **Using the network/warbler combination.** The other way to refurbish a candlestick phone is to use the network/warbler that was developed just for candlestick phones. One of these is shown in Figure 6-3.

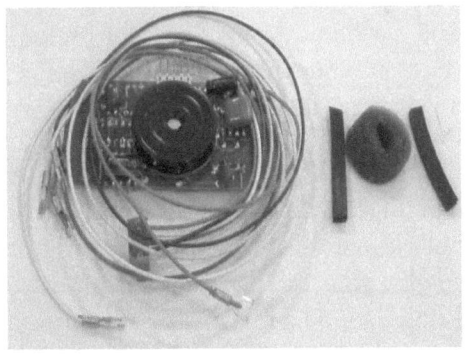

Figure 6-3. Network/warbler combination developed for candlestick phones. The large, black, round component in the center is the warbler. The transmitter is on the back side.

This has almost everything-in-one to make a candlestick work. You mount it inside the mouthpiece of an old candlestick and it includes the network, transmitter, warbler (in place of a ringer), and condenser. You use the hook switch, receiver, and dial that is already a part of the phone. In this case, there will be no ringer box. The only thing wrong with this is that the warbler has a modern electronic sound, not a real mechanical clapper banging against bells, which is characteristic of the old phones. Depending on your tastes, this is either acceptable or not, as the mechanical ringer sound is for some people a necessary part of antique phones. If this is the case for you, you can still use this network but add a mechanical ringer box that you buy from Radio Shack or one of the sources listed in Appendix J, and just plug it in series with your four-wire flat cable coming from the telephone jack. This ringer box is shown in Figure 6-4. Be sure to buy the *mechanical* ringer box with a real clapper and bells in it, not the electronic one.

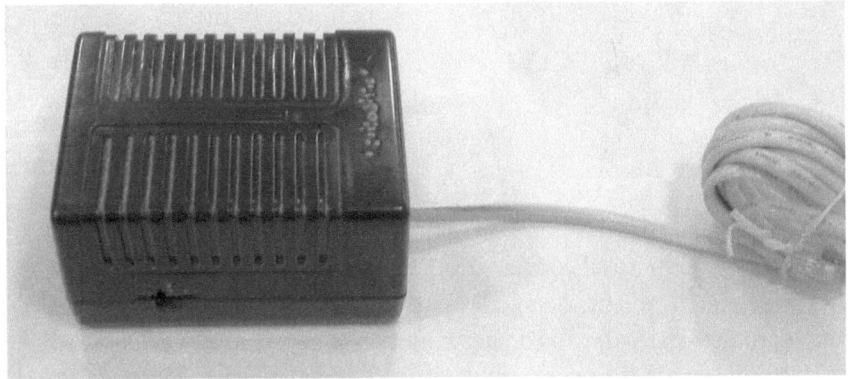

Figure 6-4. A ringer box that has a mechanical ringer in it for use on any old phone. They come in white only but this one has been spray painted black to make it more fitting for an old phone.

It is modern looking, but you can mount it against the wall down by the telephone outlet plug and paint it black like this one. To use it, you just plug it into the wall outlet and then plug your candlestick phone into the box. It produces a nice mechanical ringer sound and it completely drowns out the modern warbler sound in the candlestick. If you spray paint it black, be sure to mask off the plug using masking tape so you don't paint the connectors inside.

6.6 **Mounting and connecting the network/warbler.** The network/warbler is mounted inside the mouthpiece of the candlestick as shown in Figure 6-5. Beware that there are some candlesticks that it will not fit in, such as one with a bulldog style mouthpiece. It will fit into all the WE phones but not some of the others. So before committing yourself to using this network, make sure it will fit in the mouthpiece by taking the mouthpiece faceplate off and trying it.

Figure 6-5. Network/warbler mounted inside the mouthpiece of a candlestick phone. The network, transmitter, and warbler are all contained on the circuit board. The wires go down the stem to the base and from there connect to the line and dial. The round silver button in the center of the circuit board is the transmitter and should face out.

The wires go down inside the candlestick stem to the base and from there go to the hook switch, dial, receiver, or line. The red connector that connects the wires to the network is not keyed so it is possible to get it on backwards. Make sure you connect it on as shown in Figure 6-5 with the wires going down the back side and the transmitter, the small metal can with many small holes on the top, on the front. Fortunately, if you accidently connect it up backwards, it will not damage the network. The wiring diagram for a WE dial is shown below.

WIRING DIGRAM FOR A NETWORK/WARBLER AND A WE DIAL

- GREEN TO L2 (green wire)
- WHITE TO ONE SIDE OF THE HOOK SWITCH
- RED and L1 (red wire) TO Y ON DIAL
- BLUE TO BB ON DIAL
- YELLOW TO ONE SIDE OF THE RECEIVER
- OTHER SIDE OF RECEIVER TO W ON DIAL
- OTHER SIDE OF HOOK SWITCH TO BK ON DIAL
- THE R ON THE DIAL HAS NO CONNECTION

The wiring diagram for an AE dial is shown below. To avoid confusion, be sure to mark the connectors on the dial using pressure sensitive labels. Mark BR and Y for the impulse switch, and R and W for the shunt switch. Some of the dial connectors have more than one wire connected to them so you have to be sure that you get the right ones.

WIRING DIAGRAM FOR A NETWORK/WARBLER AND AN AE DIAL

- GREEN TO L2 (green wire)
- WHITE TO ONE SIDE OF HOOK SWITCH
- RED TO BR ON DIAL
- L1 (red wire) TO BR ON DIAL
- BLUE TO W ON DIAL
- YELLOW TO R ON DIAL
- OTHER SIDE OF HOOK SWITCH TO Y ON DIAL
- ONE SIDE OF RECEIVER TO R ON DIAL
- OTHER SIDE OF RECEIVER TO W ON DIAL

If you don't have a dial, the wiring diagram is as follows;

WIRING DIAGRAM FOR A NETWORK/WARBLER AND NO DIAL

- GREEN TO L2 (green wire)
- WHITE TO ONE SIDE OF HOOK SWITCH

- L1 (red wire) TO RED AND OTHER SIDE OF HOOK SWITCH
- BLUE TO ONE SIDE OF RECEIVER
- YELLOW TO OTHER SIDE OF RECEIVER

6.7 Using a modified mini-network in the base of a WE candlestick phone. A modified mini-network can be used in place of a network/warbler by following the directions given in paragraph 7.7 of Chapter 7. The main advantage of this is that mini-networks are about 1/3 the cost of network/warblers. The modified mini-network works with WE candlestick phones that either have dials or do not. Since a mini-network does not have a built-in transmitter like the network/warbler does, you must install one in the mouthpiece of the phone as described in paragraph E.3. Figure 6-6 shows a working WE candlestick phone with a modified mini-network in the base.

Figure 6-6. A working WE candlestick phone with a modified mini-network in the base. A candlestick phone with a dial in it could have been used as well.

6.8 Bulldog style Stromberg Carlson phone. Figure 6-7 shows a Stromberg Carlson candlestick phone with a bulldog style mouthpiece. This candlestick is notable in that:

- A network/warbler will not fit in it.
- The method of loosening the mouthpiece from the stem for adjusting the desired mouthpiece angle is different from others.

Regarding the network/warbler not fitting in, this means you can't use the network/warbler in this phone but you must use a separate ringer box as described in paragraph 6.3. Regarding the method of loosening the mouthpiece, on other brands of candlesticks there is just a bolt and nut you loosen, adjust the angle the way you want it, and then retighten the nut. With this phone you screw the knurled knob (see black arrow) on the top of the stem clockwise so it screws down on the stem. This will loosen the pressure on the swivel that the mouthpiece is attached to and you can adjust the angle to the way you want it. Then turn the knurled knob counterclockwise until the pressure is replaced on the swivel to secure it in position.

Figure 6-7. Stromberg Carlson candlestick phone with a bulldog style mouthpiece. Note the knurled knob next to the arrow which is loosened to adjust the angle of the mouthpiece.

6.9 Adding a dial to a candlestick phone that has none. There are many more candlestick phones available to purchase that do not have built-in dials than those that do. This is because the candlestick came out in the 1900's when dials were not or were very rarely used. Then in the 1930's they started to produce them with dials until the candlestick went out of style and was replaced by the models 202 and 302 phones. So most of the candlesticks you will find are non-dial types. But you can add a dial by purchasing a dial adaptor from one of the sources in Appendix J. There are three mounting holes in the back of the adaptor and you must drill three holes in the base of the phone with a #43 drill bit, tap it with a size 4-40 tap, and then put three #4-40 screws in it to secure the adapter to the phone base. Drill a much larger hole in the middle for the wires to go through such as a ¼ inch diameter hole. Figure 6-8 shows a non-dial candlestick with a dial adaptor attached. Be sure to mount the adaptor with the hook switch 90 degrees to your left.

Figure 6-8. Candlestick phone with a dial adaptor attached. The dial adaptor should be mounted with the hook switch at 90 degrees to the left.

6.10 **Using a touch tone dial with a candlestick phone.** If you want a dial with a candlestick phone but don't want to add a dial adaptor to the old candlestick desk set, you can place either a rotary dial or touch tone dial in the ringer box, along with the ringer, network, and magneto, if it has one. Whichever type of dial you choose you will need five or six wires to connect the candlestick phone to the ringer box, depending on whether you are using a rotary dial or touch tone dial. The five wires for the rotary dial are: two to the hook switch, and three to the receiver and transmitter elements (see Figure 7-3). For the touch tone dial you will need all six wires; two for the hook switch, two for the receiver, and two for the transmitter. Either way, this will require using a five foot length of six wire brown cloth covered cable, part number H2096, that can be purchased from the supplier listed in paragraph J.3 and crimp connectors from paragraph J.4. If you are using a rotary dial, follow the instructions in Figures 5-4 or 5-5, depending on whether you are using a WE or AE dial. If you are using a touch tone dial follow the instructions in Chapter 2.

6.11 **Examples of candlestick phones with the dial in the ringer box.** Figures 6-9 and 6-10 show examples of candlestick phones wired with a touch tone dial in the ringer box. Mount the ringer box up on the wall about chest high and place the desk set on your desk. To make a call, open the ringer box and punch in the number you want. If you get a computer answering system that asks for more information, just punch the information in until you finally get the person or recording you want. Then close the ringer box cover and talk while sitting at your desk just as it was used in the olden days. When you are done, you have two old antiques that are eye catching and useful. Hang them in your office, kitchen, hallway, or wherever you use a phone. They are really fun to use and appealing to visitors.

Figure 6-9. (See caption below).

Figures 6-9 and 6-10. A candlestick phone with a touch tone dial mounted in the ringer box. A rotary dial from either the 1940's or 1960's could have been installed as well.

CHAPTER 7.
REFURBISHING PHONES USING THE
MINI-NETWORK

7.1 **The mini-network.** The AE company makes a mini-network which can take the place of larger networks made by WE, AE, or any other manufacturer. It is useful when you have limited space to work with in a phone, and is shown in Figure 7-1. The only disadvantage is you cannot use it with a touch tone dial.

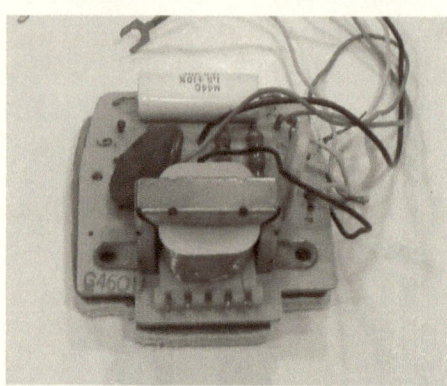

Figure 7-1, Mini-network model WA1194A. Its' small size makes it useful when there is not enough space for a WE network.

7.2 **Mounting the mini-network.** The bottom of the circuit board has exposed wiring so you must put an insulator under it when you mount it to a metal plate, usually the inside of the bottom of the phone casing. The insulator shown in the above figure is a piece of thin plywood, but any non-conductor can be used, such as a sturdy piece of cardboard, but not corrugated cardboard. There are two holes for mounting. This network is very delicate and care must be taken to protect it. On the bottom center of the transformer in Figure 7-1 you can see very thin magnet wires connecting the transformer windings to the pins in front of it. These magnet wires break very easily and they are unprotected from fingers or anything else hitting them. So be extremely careful in picking them up to insure you don't touch the wires. You can look at them with a magnifying glass to see if they have been broken. Re-soldering them is also possible, if you are lucky and good at using a soldering gun.

7.3 **Two renditions of mini-network.** There are at least two renditions of this mini-network: the model WA 1194A which is shown above, and the model WA 1236A. They both work the same way but have a slightly different method of hookup. I recommend requesting this model when ordering, but if you can't get it, be sure you ask them to enclose a wiring diagram or a conversion table for the model they are sending.

7.4 **Refurbishing a Stromberg Carlson phone.** Figure 7-2 shows a 1940's Stromberg Carlson phone that did not work so the original network was removed and a mini-network mounted in its' place. The mini-network is in the lower right hand corner. The ringer also did not work, so in the background and underneath the dial you can see a ringer salvaged from a princess telephone mounted there. Replacing two parts in the same telephone is possible only by utilizing the small sizes of the mini-network and the princess ringer. Even then, they just barely fit. Note that in the lower left hand corner near where the coiled cable comes in from the handset, a terminal strip has been added to facilitate the wiring.

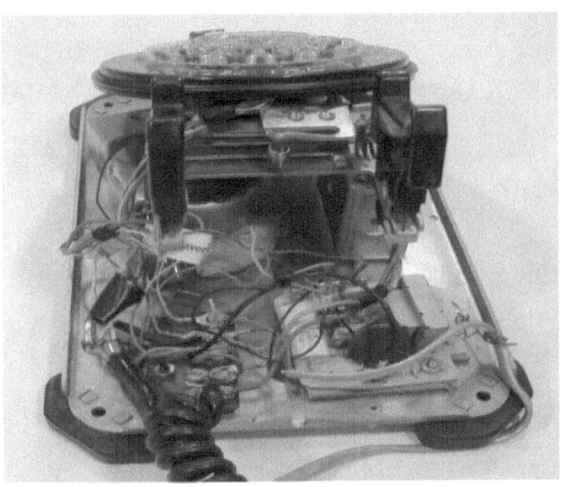

Figure 7-2. Stromberg Carlson phone with a mini-network replacing the original network (lower right hand corner) and a small ringer from a Princess phone (underneath the dial, left side) replacing the original ringer. A terminal strip was added as shown in the lower left hand corner near where the coiled cable comes in to facilitate wiring.

Also note that in the lower right hand corner, #16 gage solid non-insulted wire is used to secure the 4 wire flat line cord to a tie-down point on the phone base so it won't pull out if the phone is accidently dropped.

7.5 **Wiring diagrams for the mini-network.** It is far simpler to wire up a mini-network if you configure the handset from a 4 wire configuration to a 3 wire configuration. This is described in paragraph 5.7, and should be done prior to wiring the circuits in this chapter. Once you have the network and ringer mounted, the wiring is done according to the description below. Since the Stromberg Carlson dial is the same as an AE dial, the wiring diagram for an AE dial is given. The terminal points on an AE dial are not marked as they are on a WE dial so it is suggested that you mark them using pressure sensitive labels to avoid confusion. The terminals for an AE dial are R and W for the shunt switch, and BR and Y for the impulse switch.

WIRING DIAGRAM FOR A MINI-NETWORK AND AN AE DIAL

- LEAD 1 (YELLOW). R ON DIAL AND RECEIVER TERMINAL ON HANDSET
- LEAD 2 (WHITE). W ON DIAL AND ONE SIDE OF HOOK SWITCH
- LEAD 3 (PINK). BR ON DIAL
- LEAD 4 (BLUE). COMMON TERMINAL ON THE HANDSET
- LEAD 5 (RED). COMMON TERMINAL ON THE HANDSET
- LEAD 6 (GREEN). TRANSMITTER TERMINAL ON THE HANDSET
- L2 TO OTHER SIDE OF HOOK SWITCH
- L1 TO Y ON DIAL
- THE BLACK LEAD IS NOT USED AND MAY BE CUT

The wiring diagram for a WE dial is:

WIRING DIAGRAM FOR A MINI-NETWORK AND A WE DIAL

- LEAD 1 (YELLOW). BB ON DIAL
- LEAD 2 (WHITE) R ON DIAL AND L1
- LEAD 3 (PINK). BK ON DIAL
- LEAD 4 (BLUE). COMMON ON HANDSET
- LEAD 5 (RED). COMMON ON HANDSET
- LEAD 6 (GREEN). TRANSMITTER ON HANDSET
- ONE SIDE OF HOOK SWITCH TO Y ON DIAL
- L2 TO OTHER SIDE OF HOOK SWITCH
- L1 TO R ON DIAL
- W ON DIAL TO RECEIVER ON HANDSET
- THE BLACK LEAD IS NOT USED AND MAY BE CUT

Finally, if you don't have any dial, the wiring diagram is as follows;

WIRING DIAGRAM FOR A MINI-NETWORK WITH NO DIAL

- LEAD 1 (YELLOW). RECEIVER ON HANDSET
- LEAD 2 (WHITE). ONE SIDE OF HOOK SWITCH
- LEAD 3 (PINK). TO L1
- LEAD 4 (BLUE) COMMON ON HANDSET
- LEAD 5 (RED) COMMON ON HANDSET
- LEAD 6 (GREEN) TRANSMITTER ON HANDSET
- L2 OTHER SIDE OF HOOK SWITCH
- L1 TO PINK
- THE BLACK LEAD IS NOT USED AND MAY BE CUT

7.6 **Wiring a mini-network directly into the base of a WE model 202 phone.** The WE model 102 or 202 telephone, such as the one as shown in Figure 3-1, has a base large enough to take a mini network directly into it so you don't have to put it in an attached ringer box. It will have everything in it except a ringer but you can attach a separate mechanical ringer such as the one shown in Figure 6-4 to have a completed phone. There is a real shortage of space in the phone so is not easy to do, but it is possible, particularly if you do two things while assembling it. They are:

- Use short wires when wiring it up so you will not have long wiring bundles that you have to find a place for when you put the base on.
- Get rid of the connection where the "R" connection is located. This will open up an area for the transformer in the mini network to go. The transformer is the largest component in the network.

To do this, refer to Figure 7-3. The "R" connection mount located at the end of the black arrow on the

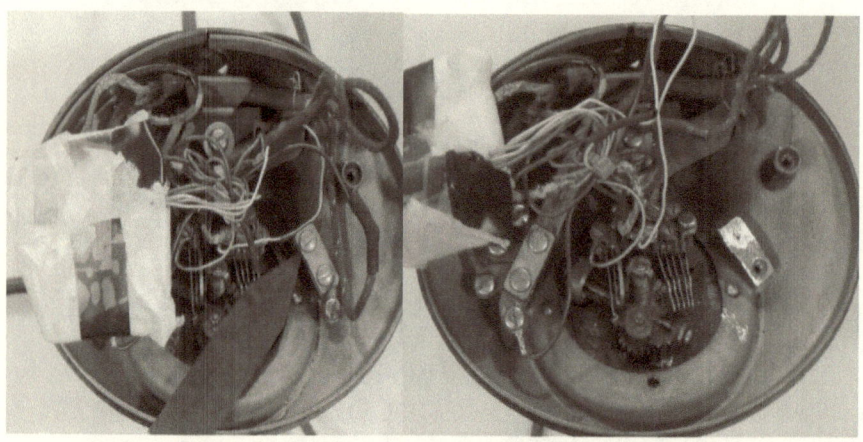

Figure 7-3 (Left). The arrow points to the "R" connector that should be removed.

Figure 7-4 (Right). The connecter after it has been removed with a screw driver and the shoulder of the mounting base has been ground down.

right side is not used when you are using a mini network so this can be eliminated to make more space. Unscrew the two screws holding the "R" connector, and then grind down the corner of the mounting base as much as you can using a grinder stone attached to a drill, as shown in Figure 7-5.

Figure 7-5. Grinder attached to an electrical powered drill used to grind down the shoulder of the "R" connector.

Use an electrical powered drill, not a battery powered one as it will take quite a bit of grinding and the batteries of a battery powered one would never stand up to it. Figure 7-4 shows the "R" connecter ground down. When the wiring of the mini network is done and tested, mount the network with the transformer in the pit created by eliminating the "R" terminal. Figure 7-6 shows the mini network in its final position. The hook switch plunger is free to move both up and down with no wires on either side and the dial has no wires interfering with either the spindle or leaf springs.

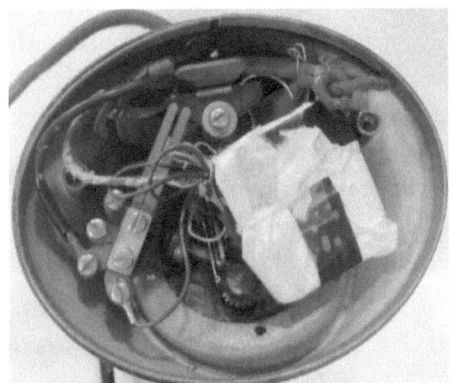

Figure 7-6. The final position for the mini network. Notice that the hook switch plunger has no wires around it so it is free to move up and down. Also, the dial spindle and dial leaf springs are unencumbered, although that can not be seen in this figure.

Wrap the mini-network with paper towel and then electrical tape so nothing will short it out.

7.7 **Modifying the mini-network to reduce its' size.** The mini-network can easily be reduced in size to allow it to be used in the base of a WE candlestick phone, a WE model 202 phone such as the one described above in paragraph 7.6, or some other location that is space limited such as in a European phone or an old intercom. Figure 7-7 shows a mini-network that is unmodified and one that is modified.

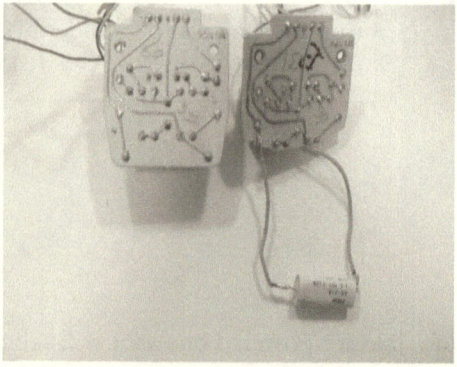

Figure 7-7. An unmodified mini-network on the left and a modified one to reduce its size on the right.

To do this, remove the large capacitor at the opposite end from the transformer by heating the solder at each end with a soldering gun and pulling it off. Then, using a coping saw, cut the circuit board to make it as short as you can without removing any other circuit components. Then solder the capacitor back on but using two pieces of wire about 2 1/2 inches long. This will reduce the size significantly and you can put the capacitor anywhere you can fit it in. This modification takes about 15 minutes but greatly increases the options of where you can use it in a space restricted phone, which many of them are. The network should be wrapped with electrical tape to prevent any short circuits with other components or the phone case. Before putting electrical tape on, be sure to put a small piece of paper towel over the small wires coming out of the transformer that connect to the binding posts. This will prevent them from being pulled off when you remove the tape. Also, since the mini-network does not have a built-in transmitter like the network/warbler does, you must install one in the mouthpiece of the phone.

CHAPTER 8.
CONVERTING AN OLD INTERCOM TO A TELEPHONE

8.1 **Old intercoms.** Any old intercom can be converted to a telephone. They are still available in antique stores because they were very common 100 years ago and were used extensively in factories and stores. They are mostly wall mounted and they make very interesting display items on the wall as well as useful telephones. You can doll them up with walnut mounting bases and they can look very elegant on a home or office wall.

8.2 **The insides of an intercom.** The electronics in intercoms is totally different than telephones in that they worked on DC batteries located in the home or office instead of AC power from the phone company's switch room. Thus, the transmitter, receiver, network, and ringer will not work when you convert them to telephones so you have to take those items out and put in replacements that are compatible with telephones. In short, to convert an intercom to a phone, you keep the handset, hook switch, and wall housing and discard everything else. Figure 8-1 shows an intercom from the 1920's that will be converted into a handsome looking useable wall phone. Figure 8-2 shows the insides of the handset.

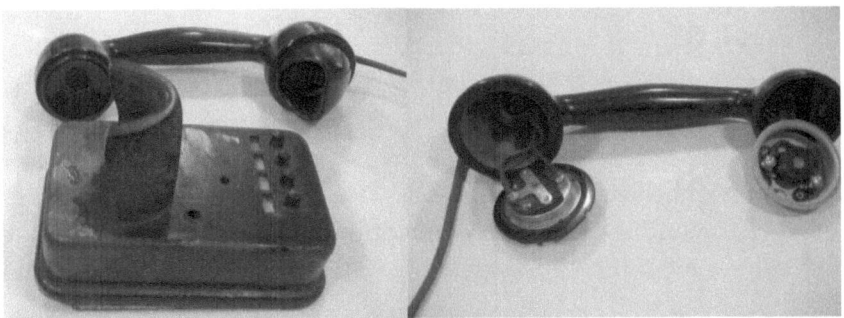

Left. Figure 8-1. Old wall mounted intercom that will be converted into a telephone. Right. Figure 8-2. The handset with the insides of the receiver and transmitter exposed.

Note in Figure 8-1 that the buttons for calling different stations are there along with places to write what stations each one was. Before painting, it is best to remove and save the old paper and then replace it when done painting. If it has the old stations still marked on it, this makes a very interesting conversation piece. Figure 8-2 shows the original receiver and transmitter elements that are not compatible with telephones so they must be discarded and replaced with compatible ones. This is easily done as you can use the old original wiring with the replacement elements.

8.3 **Removing the insides.** Figure 8-3 shows the insides of the intercom. The ringer, and other electronics must be taken out and discarded but be sure you keep the working hook switch. Also, keep the station select buttons but disconnect all the wiring to them as they will no longer apply.

Figure 8-3. Insides of the intercom. All except the working hook switch and the station select switch mechanism should be removed.

You are pretty much on your own in removing the insides as every one is different, but just make sure the phone body has a working hook switch and that the station select switch mechanism is there and also the original station markers to add authenticity and interest.

8.4 **Preparing the intercom body.** For this phone a mini-network was installed as space was at a premium. The mini-network is installed using the wiring diagram shown in Chapter 7, and the warbler is installed as shown in Appendix D. A walnut mounting base was made and a warbler was mounted below the intercom. The wires to the warbler are routed in back. The mounting base has three pre-drilled holes in the corners to make it easy for the end user to mount to the wall.

8.5 **The finished telephone.** Figure 8-4 and 8-5 show the finished intercom. Figure 8-4 shows the insides after wiring and Figure 8-5 the completed unit.

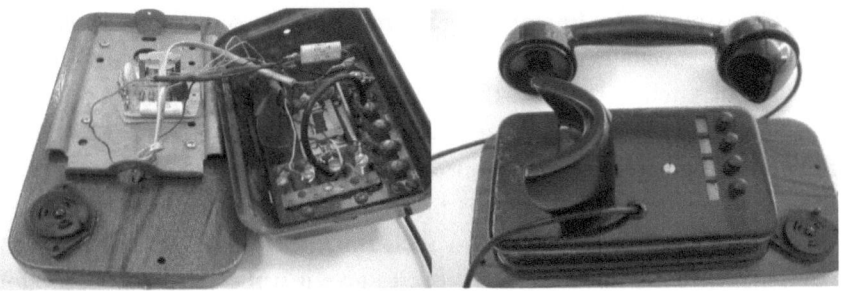

Left. Figure 8-4. The completed converted intercom opened up to show the wiring of the mini-network to the hook switch, receiver, and transmitter.

Right. Figure 8-5. The completed unit mounted on a piece of solid walnut

There was no room for a ringer inside so a warbler was mounted on the walnut base and the wiring run along the back. The warbler was painted black to make it aesthetically compatible with an old phone look. The result is a handsome looking extension phone, (meaning it sends, receives, rings, but no dial), that is useful and an interesting remnant from the past.

8.6 **Using an exposed ringer.** A variation from the above intercom is one shown in Figure 8-6. This one is basically the same as far as converting it to a telephone is concerned, but it is a different intercom and has a mechanical ringer attached instead of a warbler. It is shown in Figure 8-6.

Figure 8-6. An intercom with a mechanical ringer from a 1970's princess phone. It is called an extension phone as it sends, receives, rings, but has no dial.

It is a good looking and interesting wall mounted extension phone that is useable. People like to see the ringer exposed and to see a 100 year old intercom.

8.7 Using a false bottom to gain more room. A lot of the intercoms are so full of mechanical apparatus used for switching in and switching out the various satellite stations, that there just isn't enough room to get the network in. One way of getting enough room is to make a false bottom in the mounting base. Figure 8-7 shows the bottom of an old intercom with 8 satellite stations and there was no room for a mini-network or ringer. So a wood mounting base of 3/4 inch thickness was used and the part covered by the phone was routed out to give it sufficient room to mount both the network and ringer. It is desirable to set the depth of the router bit so that a small amount of wood is left to act as a base. This will protect the electronics inside the phone. If you have to route out the entire volume, then glue a black, brown, or green felt bottom on after you have it routed out. Another way of getting more room is to utilize the modified mini-network as discussed in paragraph 7.7 of Chapter 7.

8.8 Mounting the warbler. The warbler should be countersunk into the backboard to keep it from protruding out too far. To do that, use a 1 3/8 inch diameter hole saw as shown in Figure 8-8. The resulting hole is just a little too small for the warbler to fit into, so use a rasp bit connected to a drill to make it little bit bigger. The warbler will then fit in snugly. For the electrical hookup, the warbler must be connected in series with a 0.46 microfarad or larger capacitor as shown in Figure 8-9. If you are connecting the warbler to a magneto instead of the telephone line, it is not necessary to use a capacitor.

Figure 8-7 False bottom allowing most of electronics to be buried in the base of the phone.

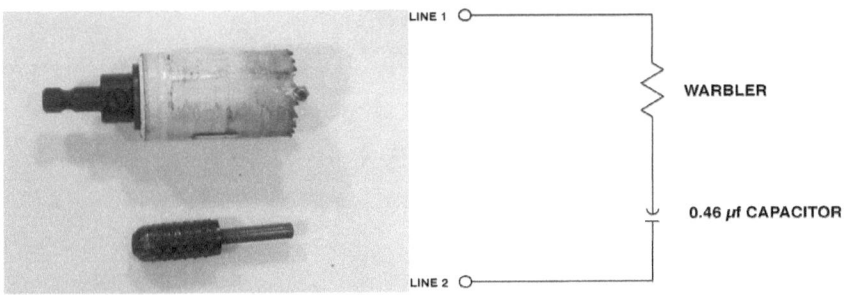

Figure 8-8 (left). Hole saw and rasp for mounting warbler.

Figure 8-9 (right). Diagram for wiring warbler to a telephone line.

CHAPTER 9.
TROUBLESHOOTING

9.1 **Preventive steps.** When you finish wiring a phone and it does not work, don't feel badly as that is usually the case. There are lots of wire connections to make and every one has to be correct or you don't have a working phone. Hopefully, you have taken the preventive steps recommended in earlier chapters to minimize the chance of an error. Those are reviewed here

- You should always wire up the ringer circuit first, then plug it in to a phone line and call it to see if you have that part wired correctly.
- Whenever you crimp a connector onto a wire, you should check it with a volt-ohm meter to make sure the connection at both ends is good. Open wires are the hardest problems to detect because you can't see them.
- If you are using a network other than a mini-network, a common error is neglecting to jump L1 to RR.

9.2 **Checking the connection to the new electronics.** First, plug the phone into a line and, with the hook switch hung up (pressed down), call the line from another phone and see if it rings. If not, it should be easy to fix as something is wrong with the line up to the ringer and hook switch. This should not be the case as you have already tested this part, but things can go wrong while working on the rest of the phone. Then, while the line is still ringing, let the hook switch come ON and see if the ringer stops ringing. If it doesn't, the main body of the phone is not connected. The circuits in the switch room measure the load on the line to know when to turn off the ring signal. If there is no load when the hook switch is ON, something is wired wrong from the hook switch to the rest of the phone wiring. Check the wiring from the hook switch to the dial and from the dial to the network.

9.3 Checking for a dial tone. Next, see if you have a dial tone. If there is no dial tone, the problem is harder to fix, as it could be in the receiver circuit meaning the dial tone is really there but the receiver is not working, or in the input connections to the network. Re-check the receiver connections. If they look OK, start checking further into the connections to the dial and those to the network. From here, you just have to keep looking, as the problem could be anywhere. Listen to the receiver while flipping the hook switch up and down. If you hear a faint clicking sound, the receiver is probably OK, but the problem is somewhere in the connections to the dial or network. Listen to the receiver while turning the dial and letting it wind down. If you hear clicking sounds, the problem might be in the shunt circuit wiring or the receiver but probably not in the impulse circuit.

9.4 Tracing the dial tone to the impulse switch. At some point, it is helpful to use a volt ohm meter to measure AC volts at various places in the wiring. First, set the meter to AC volts on the 10 volts scale, or a scale close to that, and put the probes on the terminals where the power cable comes in, that is, L1 and L2. You should get some kind of a reading. It doesn't matter what the reading is, as long as you are getting something. This proves that your connections to the outside telephone line are OK so far, as the signal you are reading is the dial tone. Then, you have to use your own reasoning to follow the wires to their next point and measure there. From the incoming terminal, the signal will go (on one of the lines) to the impulse switch, so check for the signal there. This is the general approach and you have to look at the wiring diagram to figure out where to measure next.

9.5 Checking the dial pulses. Once you do get a dial tone, the problem is a lot easier to trace. Dial a known number such as another number in the house to see if the dial works. If you don't have another known number to dial, dial the line you are calling from. You should get a busy signal. If not, check the wiring from both sides of the impulse switch. When you dial the first digit of a number, the dial tone sent by the switch room should go away. If it doesn't, the dial pulses are not getting sent down the line to the switch room. Check the related wiring and see if the impulse switch is opening and closing during wind-down. If the problem is that you can receive but not transmit voice signals, check the transmitter circuit which is shown if Figure 9-1.

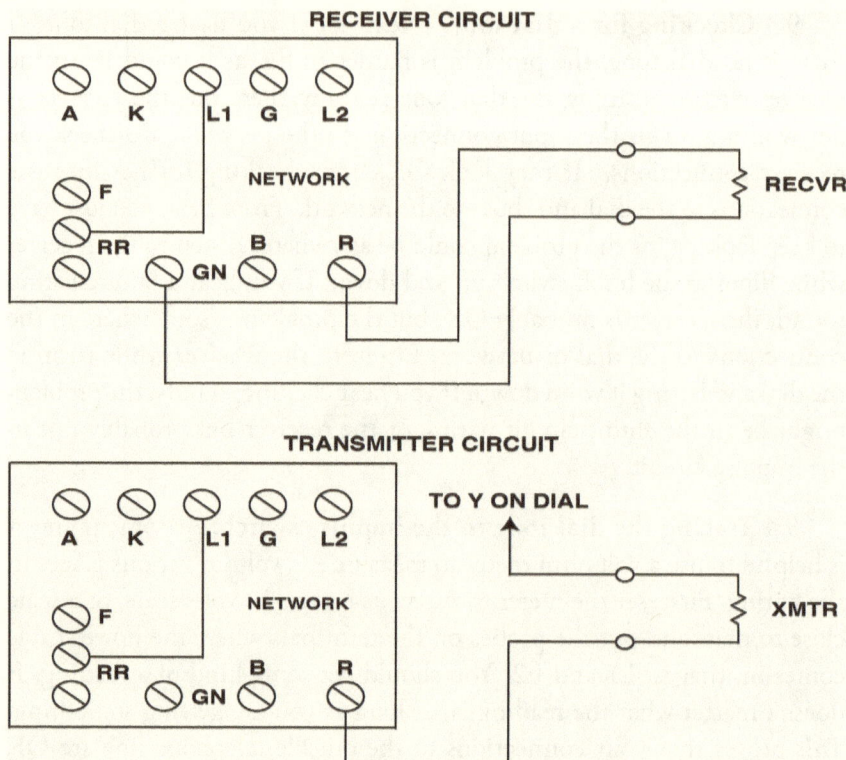

Figure 9-1. Receiver circuit and transmitter circuit simplified for use in troubleshooting. If the receiver or transmitter do not work, try looking in the above areas of the circuit to find a wiring error.

As you can see, to diagnose the problem, you try to guess where the problem is by diagnosing the type of failure, then looking in that part of the circuit. If you still can't find the problem, you can check each of the components of the phone by using the test platform shown in Appendix A.

CHAPTER 10.
REFURBISHING EUROPEAN PHONES

10.1 **Determining L1 and L2.** European phones need a separate chapter to demonstrate how to refurbish them because the handsets are much smaller and the electronics in the desk sets are much more complicated that their comparable American made ones. Additionally, there is no way to get a wiring diagram for them so they have to either work or the electronics must be totally replaced. Hopefully, you will be able to figure out where L1 and L2 are so you can try the phone out using a surface mount modular jack (see paragraph 1.8). Usually, they are marked but if they are not, you can usually figure it out if the original line cord is still connected. There can be up to five wires in the original line cord, any two of which are L1 and L2. You can try connecting two at a time to the phone line in an attempt to come across them by accident. This might take a while but it's easier than replacing the whole system. If you succeed, there is a good chance the phone will work except for maybe needing to replace the receiver, transmitter, handset cord, or all three. The line cord will certainly need replacing as it needs the modern modular connector on one end.

10.2 **Salvaging the original electronics.** Figure 10-1 shows an Ericsson desk set with the cover and bell removed. When taking apart any of the European phones *be sure to save all the bolts and nuts as they are metric and will be very hard to replace, even if your hardware store has a wide selection of metric hardware.* Many of the nuts and bolts in Europe in those days were custom made for the telephone manufacturer and cannot be replaced.

Figure 10-1. Ericsson phone with the cover and bell removed. The coil is along the bottom under the metal band, the ringer coils along the top under the band, the hook switch on the top of the metal band, and the line cord and handset cord going off to the left.

If you cannot salvage the original electronics, you must replace them with a network or mini-network as shown in Figure 10-2. Beware when removing the old electronics, you must be sure that a working hook switch is left in place. In Figure 10-2, all the existing electronics except the hook switch and ringer have been removed and replaced by a model 4228 network. It is ready to wire up the phone according to the wiring diagrams as given in Chapter 5 or Chapter 7.

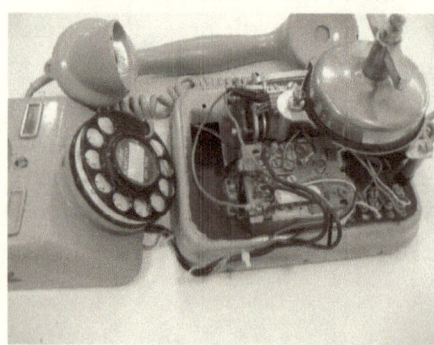

Figure 10-2. Ericsson phone with the electronics replaced with a model 4228 network. Be sure to leave a working hook switch.

10.3 **Push-to-talk switch.** The handset should be tried to see if the transmitter and receiver need replacing. This can be done in its existing setting if the electronics works, or you can disconnect the handset from the desk part and test it in a test platform. Figure10-3 shows the handset of an Ericsson with the transmitter and receiver caps off and the transmitter element removed. Notice that this one has a push-to-talk switch in the middle of the handset. Some do not but if yours does you will probably want to bypass it as no one wants to talk with a "push to talk" switch.

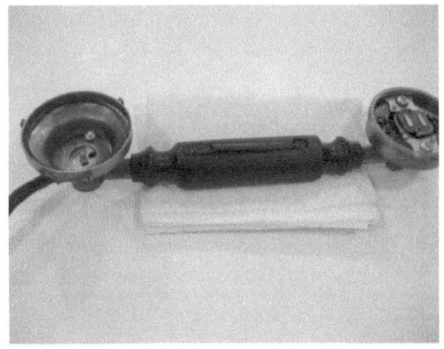

Figure 10-3. Ericsson handset with transmitter, transmitter element, and receiver caps removed. This one has a push-to-talk switch in the stem.

If you want to bypass the push-to-talk switch, you might as well replace the old receiver element also as the receiver housing has to be removed to get the wires through. Figure 10-4 shows the handset stem after the transmitter and receiver cups have been removed.

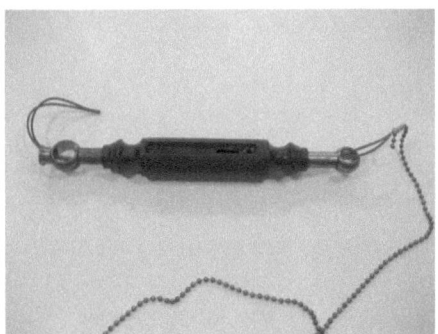

Figure 10-4. Handset stem with two wires pulled through using ball chain available at any hardware store.

When replacing the receiver element in this Ericsson and other European phones, the space in the earpiece part is so small that you will need a receiver element that is only ½ inch thick. Audiosears Corporation makes the model 2463V receiver element that has these dimensions and you can order it from the manufacturer listed in J.10. Unfortunately, they have a minimum order requirement but it is not too high. Perhaps you can share an order with someone else. The mouthpiece in an Ericsson handset is also very small so you should order a model N-1 transmitter element, which is the smallest transmitter there is. It can be purchased from the supplier listed in paragraph J.3.

10.4 **Re-wiring the handset.** Remove the receiver element and the cup and the transmitter cup and unscrew the cover that houses the push-to-talk switch. Disconnect the wires from the switch and replace the switch cover and its screws. Take two strands of hookup wire, strip off about ¾ inch of insulation on one end of each, and wrap both around one end of some ball chain, available at any hardware store. See Figure 10-4 for a depiction of this. Wrap the wire tightly so you can pull on the ball chain and the wires will not pull off. The ball chain is limp and has weight, so you can take advantage of gravity to drop it down the hole in the stem and use it to pull through the two hookup wires. Then, reconnect the transmitter and receiver housing cups. Install new transmitter and receiver elements, reconnect the handset cable, and reconnect it to the desk set.

CHAPTER 11.
REFURBISHING US ARMY FIELD
TELEPHONES MODEL EE-8

11.1 Possible failure modes of field telephones. The model EE-8 field telephone has been around since WWII. When not in a combat zone, they are used in training six days per week and sometimes 16 hours per day. Thus, although they are made to military specifications for ruggedness, they are usually worn out when they appear on the civilian market. Figure 11-1 shows a pair of field telephones.

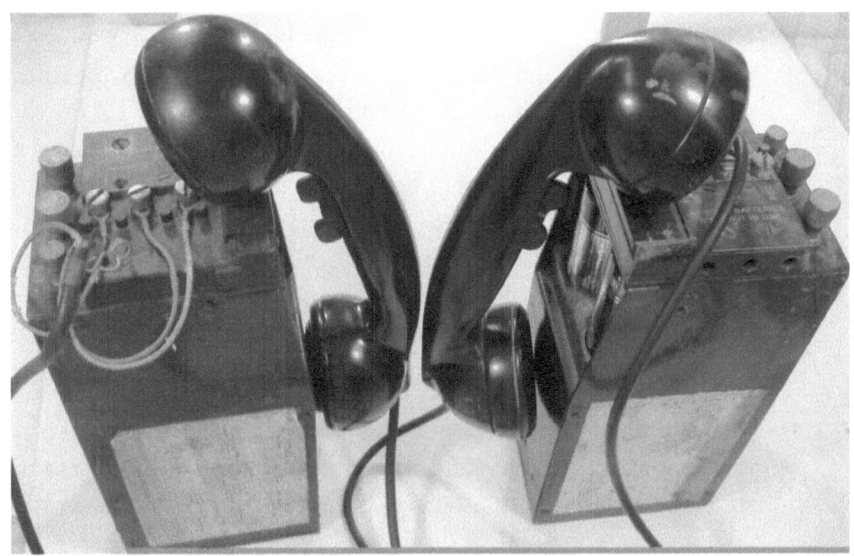

Figure 11-1. Pair of US Army model EE-8 field telephones.

These are really not telephones at all but intercoms. They cannot be connected to a telephone line and they only work with each other. Thus, there is no way to replace the electronics with a mini-network or anything else as has been done with all the other phones in this book. To refurbish one usually consists of one or more of the following:

- Repair a broken wire in the electronics box
- Replace the transmitter and/or receiver elements
- Replace the cable from the electronics box to the handset
- Replace the ringer with a warbler

Each of these possibilities is discussed separately.

- 11.2 **Repair a broken wire in the electronics box.** First look to see if there are any obviously broken wires and reattach them if there are. When that is done, test the electronics box by inserting known good D cell batteries and measure the voltage from the BAT − terminal to the BAT + terminal using a volt ohm meter. The voltage should be 3 volts DC. If it isn't, make sure the terminals where the batteries are installed are not corroded. Sand them down with fine sandpaper such as #220 grit to get off any corrosion. When you do measure 3 volts, try hooking the two phones together and see if they work. To hook them together, simply connect L1 and L2 on one phone to L1 and L2 on the other phone. You should hear your voice in the others' receiver. Try this for the other phone and you should again hear your voice in the other's receiver. Note that you have to rotate the butterfly switch on each handset to talk. The phones will work over any distance that you can connect a pair of wires. There is a potentiometer adjust on the top of each phone. This should be set to the maximum counter-clockwise (ccw) position.

- 11.3 **Replace the transmitter or receiver elements**. If the transmitter and/or receiver elements fail, they are easily replaceable since the handset is the same as those used in the manufactures civilian telephone. This is discussed in Appendix E.

- 11.4 **Replace the cable from the electronics box to the handset**. If the cables are worn and frayed which is usually the case, they can easily be replaced. Figure 11-2 shows a detail of the top of the phone where the connections are made.

Figure 11-2. Top view of an EE-8 Army field telephone showing all the connectors.

The three connectors along the right side arranged vertically are labeled L1, L2, and BAT -, from top to bottom. The four connectors in the middle arranged from right to left are labeled L2, REC, C, and T & BAT +. In this case, the green, red, and yellow cables are badly worn and must be replaced. The same cable was used but the black insulation was cut back a few inches exposing new-looking red, yellow, and green cables that were in very good shape. The old worn out cables were cut and new connectors crimped on to make a near new cable as shown in Figure 11-3.

Figure 11-3. Old handset cable with 2 inches of black outer sheath cut off and the resulting wires used for connecting to the terminals.

11.5 **Replace the ringer with a warbler.** If a ringer does not work when you turn the magneto crank on another phone connected to this one, take the side panel off the non-working one as shown in Figure 11-4. The ringer is located in the bottom compartment so you can disconnect the ringer and measure the voltage on the two wires leading to it while someone is turning the crank of the other. The voltage should be roughly

71

50 volts as measured on an AC voltage setting. If the voltage is reduced to as much as 30 volts, you can replace the mechanical ringer with a warbler and it will probably work fine. This was done to the phone in figure 11-3. The black cylinder with the red wire going to it is a replacement warbler and easily fits into the compartment where the old ringer was.

Figure 11-4. US Army field telephone with side panel removed, showing replacement warbler in the ringer compartment, bottom, with red wire going to it.

11.6 **Using old Army telephones.** Old Army field telephones are fun to use, interesting to look at, and will work anywhere you can string a pair of wires. The D cells will last a long time because the push-to-talk switch limits every user's length of conversation.

CHAPTER 12.
SERIES CONNECTION OF TELEPHONES

12.1 **What the series connection is.** It is possible to make a telephone without using a network, but using only a dial, receiver, and transmitter. The problem is that sometimes they don't work, or if they work, the volume is so low that they are not a viable telephone. The design is called a series connection because the hook switch, impulse switch, receiver, and transmitter are all connected in series and then connected to L1 and L2 without any coil being involved. The only advantage of using this circuit is where you don't have enough room to fit a network in. This series connection is only included in this book to make it as complete a collection of telephone circuit diagrams as possible and to give you the option of using it if absolutely necessary. But use of this circuit is not recommended.

12.2 **Series connection circuit diagrams.** Figures 12-1 on the left and 12-2 on the right show the circuit diagram for the series connection for both the WE dial and AE dial.

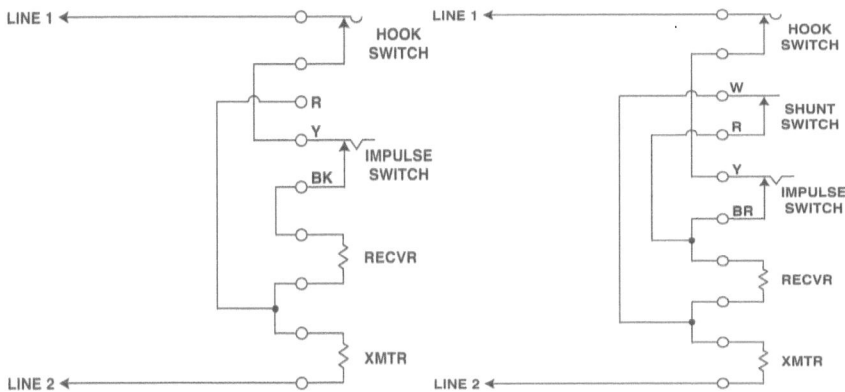

Figures 12-1 and 12-2. The series connection for telephones using the WE dial and the AE dial.

12.3 Alex Bell didn't use it. As you can see, it is a simple series connection with the shunt switch used to block out the dialing pulses in the receiver for the AE case, and the "R" connection used to block them out in the WE case. Again, it is here for your reference and knowledge but it is not recommended for use. Alex Bell didn't use it when he invented the telephone in 1875, and for good reason.

CHAPTER 13.
REFURBISHING PAY TELEPHONES

13.1 **The electronics in pay telephones.** Pay telephones use the same electronics as regular phones. The method of getting and counting the coins is a side circuit and is not used when you refurbish them for use today. So you should be able to use the same circuitry as is there already. They are easy to refurbish, but you need to make a wooden back plate that is a little larger than the phone itself so it is convenient and easy for the end user to attach it to the wall.

13.2 **Getting in to the electronics box.** Pay phones needed to be burglar proof so they were made of thick rolled steel which makes them very heavy and rugged. There are two locks: one for the coin box and one for the electronics box. You do not need to get into the coin box but you can't refurbish them unless you can get into the electronics box. So before you buy one, be sure you can break into the electronics box; either by opening it with a key, drilling the lock out, or unscrewing the back. The ones made in the 1950's and later are easier as all you have to do is take the back off by undoing some screws. The earlier ones have a steel back and a lock in the bottom. If that lock is still intact and you don't have the key, you have to drill the lock out to open it. This can be some work but it is doable.

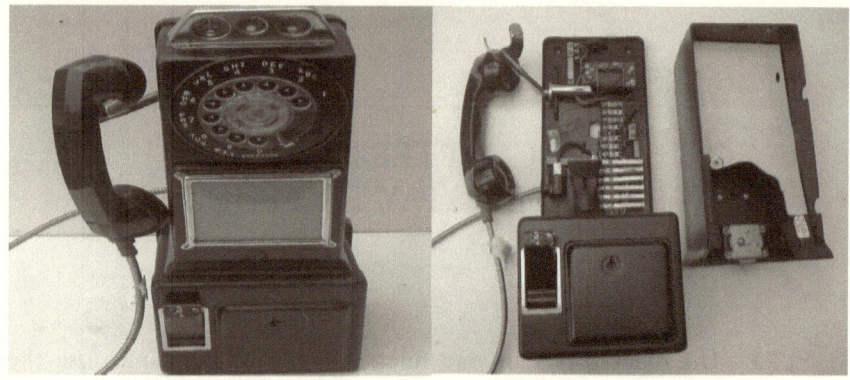

Figure 13-1 (Left) 1950's pay phone.

Figure 13-2 (Right) Pay phone with the electronics box opened

13.3 **Refurbishing the electronics.** Figure 13.1 shows a 1960's era pay telephone and Figure 13.2 shows the insides of the electronics box. In this one near the top, you can see the coil, in this case a WE model 101A coil or similar coil that is used. Try to salvage the existing electronics. In this one, everything looked OK but there was no indication of which screw tie-downs were L1 and L2. Thus, the instructions given in paragraph 10.1 were used to determine whether the existing electronics were salvageable. The L1 and L2 lines were cut but the old spade lugs were still connected to the tie-down connectors so it was easy to tell which ones they were. If you cannot salvage the original electronics, you must replace them using the instructions given in either Chapters 3, 5, or 7. This should be easy to do as there is plenty of room in the box., but there is a very high chance that the original electronics still works, so try to salvage them.

13.4 **Testing the phone.** In Figure 13.2, you can see several spring leaf connectors that connect to other connectors when the back is mounted. These, in part, provide an automatic disconnect of the phone from the line when the back is removed. So to test it, you must make the connections you want and then re-attach the back before testing it on the phone line. Use the instructions given in Chapter 10 to try to salvage the existing electronics. If you cannot salvage the existing electronics, then you must tear out the old wiring, insert a network, and re-wire it using the wiring instructions given in either Chapters 3, 5, or 7, depending on which type of network you use. This is not as difficult a job as it sounds because all the components of a telephone, the receiver, transmitter, hook switch, dial, and network are all there. You just have to wire them together.

13.5 **The coin system.** The coin system is not used in refurbished pay telephones, so it is best to make baffles to route all the coins directly to the coin return box. Use thin sheets of aluminum that can be purchased at any hardware or hobby store. Cut and bend them to route the coins to the return box and glue them on with epoxy.

13.6 **The ringer.** Starting in the 1960's the phone companies started omitting a ringer from some pay phones, figuring they were used for making outgoing calls only. So your phone may or may not have a ringer. The one in Figure 13-1 did not. So if yours does not, you must install either a mechanical one or a warbler. The phone in Figure 13-3 has a warbler installed and it can be seen on the lower left-hand side. A hole saw like the one shown in Figure 8-8 can be used to make the hole even though the wall is solid steel. The electrical wiring diagram for wiring up a warbler is shown in Figure 8-9.

13.7 **The back mounting plate.** Pay phones can only be mounted to a wall if you have a special mounting plate that they usually do not come with. So, during the refurbishing process, a wooden pack plate for mounting it should be made. Figure 13-3 shows such a back plate that has a screw hole drilled in each corner, so it is easy for anyone to mount to a wall.

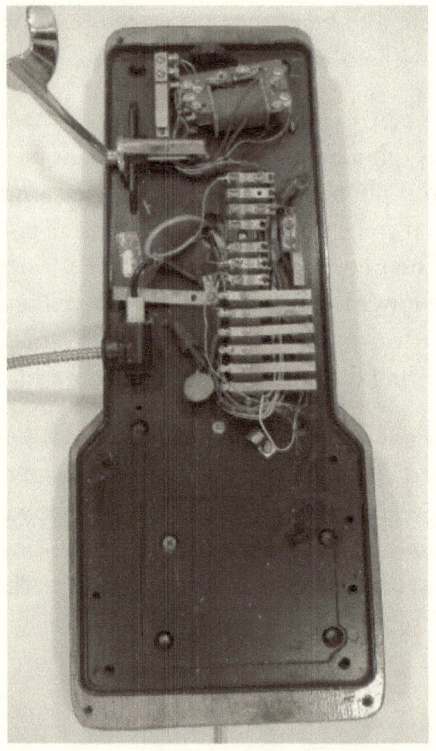

Figure 13-3. Back plate of pay telephone mounted to a wooden mounting base. This should be used for any wall mounted phone.

APPENDIX A.
BUILDING A TEST PLATFORM FOR TESTING ALL THE COMPONENTS OF A ROTARY DIAL TELEPHONE

A.1 **The components that need to be tested.** When testing a rotary dial telephone, the components you need to test are;

- Transmitter
- Receiver
- Hook switch
- Ringer
- Rotary dial

You really can't test the network as there is no easy way, and if all the other components work and the phone does not, you know it's the network. The networks usually don't fail because they are located in the protective shell of the phone body and there are no moving parts. So just go on the theory that the network is OK, unless it looks like it's been abused. On the mini-network, look at the small wires coming from the transformer to the tie-down pins with a magnifying glass to see if they are broken. If not, it is probably OK. If there are any broken wires, sometimes it is possible to solder them back together if you are adept at soldering small parts.

A.2 **Parts needed to build the platform.** You can simply and easily build a test platform for testing all the components of a phone. All you need is a WE network or a mini-network, a handset from a 1940's through 1980's telephone, a warbler, and a hook switch. All these except the mini-network and warbler can be obtained from a 1960's or 1970's phone that you don't mind taking apart. Figure A-1 shows such a test platform.

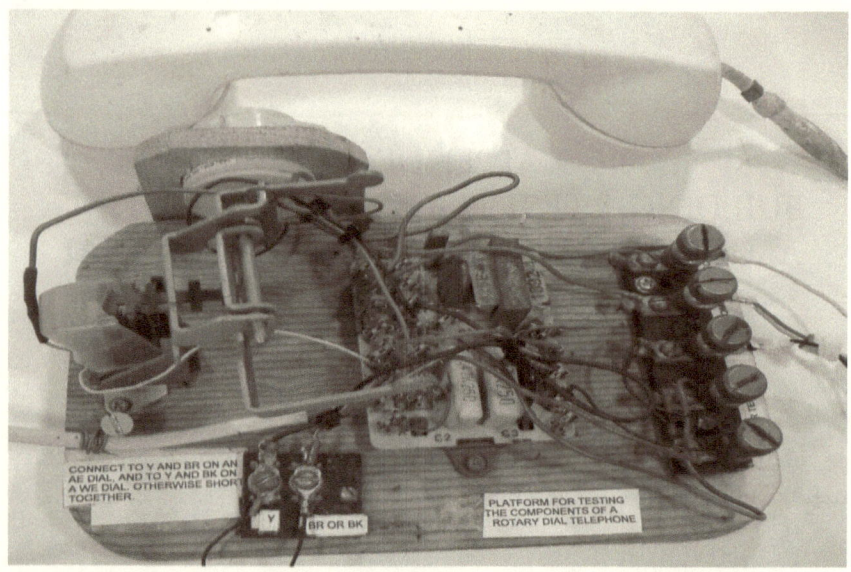

Figure A-1. Platform for testing the components of a dial telephone.

As you can see, the wooden platform has a hook switch taken from an old 1960's WE phone (on the left center), an electronic warbler on the top left, a WE network in the center, five screw-down terminals on the right, and two screw-down terminals on the bottom left. The screw-down terminals came from an old telephone but you can use a new terminal strip from Radio Shack. A model G1 hand set from a 1950's 60's or 70's phone is connected the terminals on the right. There is a line cord used to connect the platform to the telephone line on the left. As the platform stands as shown, it is a working telephone without a dial. You can call the line it is connected to and it will ring and send and receive. Remember to hold the hook switch down when you dial the test platform or you will just get a busy signal.

A.3 **Wiring diagram for the platform.** Figure A-2 shows the wiring diagram for the platform. The warbler on the test platform is optional but it makes it a little more convenient to have one when you call the platform and the phone under test does not have a ringer. Also, it is best to use an electronic warbler instead of a mechanical ringer. It takes almost no power and does not overload the circuit when you have a mechanical ringer under test.

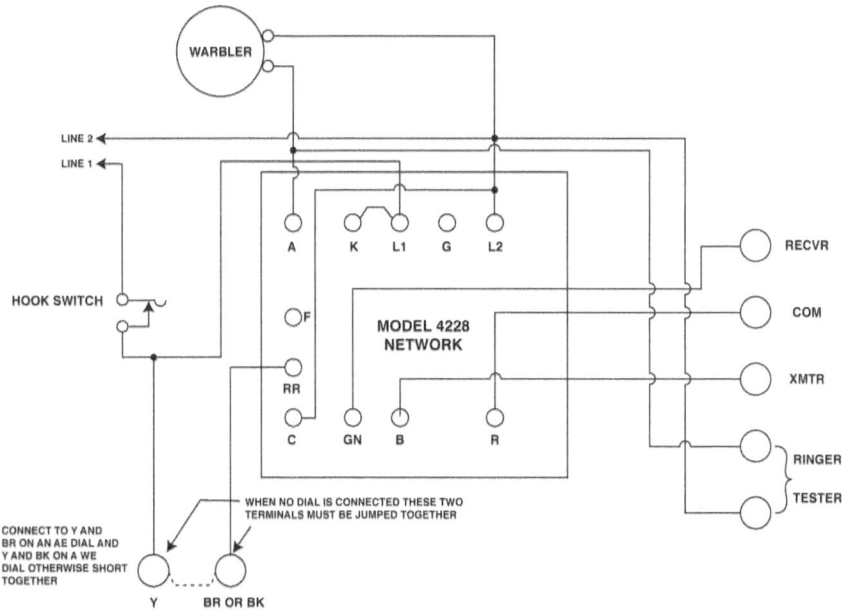

Figure A-2. Wiring diagram for test platform using a model 4228 network. The warbler is not necessary but is convenient to have when the phone you are testing does not have a working ringer.

The five terminal posts on the right and the two on the bottom are screw-down terminal posts. See Figure A-1. If your handset you are testing has two wires coming from the transmitter and two from the receiver, just combine one from each and put them in the terminal labeled "common". Figure A-3 shows the wiring diagram for the test platform using a mini-network. Please note that the 0.46 microfarad capacitor can be a larger capacitor, but not a smaller one. Don't forget to jump K to L1and C to L2.

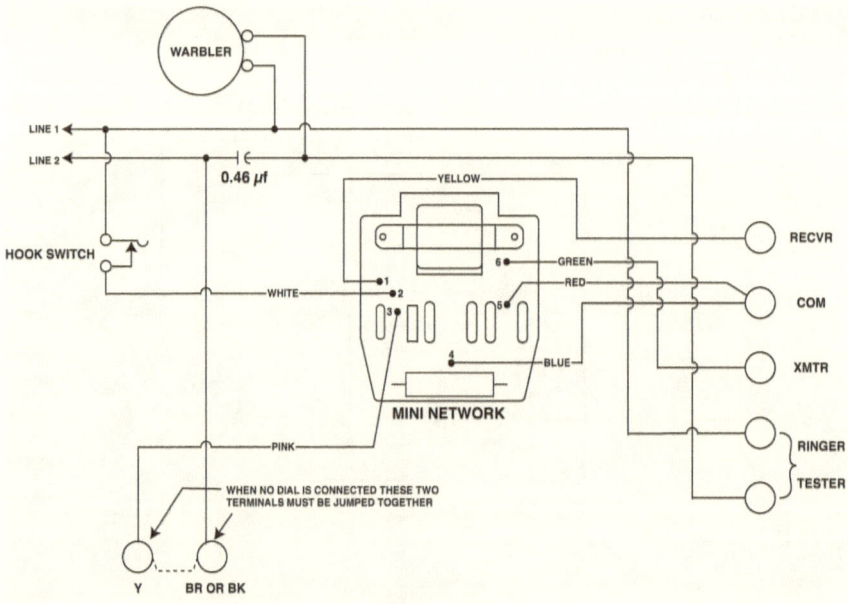

Figure A-3 Test diagram using mini-network.

A.4 **Testing a rotary dial.** The terminals on the bottom left, one says Y, and the other says BR or BK. To test a dial, you connect the dial connections to those terminals, depending on whether it's an AE dial (Y and BR) or WE dial (Y and BK) dial. Then dial a number you know is working, such as another line in your house, and see if it rings. If so, you know you have a good dial regarding the impulse side. To test the shunt side, just connect a volt-ohm meter to the two shunt terminals, set the volt-ohm meter on ohms, and see if the shunt switch behaves as shown in the timing diagram in Figure C-4 in Appendix C. Note that it is different for the WE dial and the AE dial. For the WE dial, the shunt switch should be shorted during wind up and wind down, and for the AE dial it should be open for the wind up and wind down.

A.5 **Testing a transmitter or receiver element.** Disconnect the transmitter on the test platform and connect up the one being tested. Call up the line and test it. The same for the receiver, but be sure you have at least one transmitter and one receiver connected at all times. Don't leave one disconnected.

A.6 **Testing a ringer.** Connect the ringer to the terminals that say "ringer" on the right side. Dial the line that the test platform is connected to and see if the ringer works. Again, be sure the hook switch is closed before attempting to dial the test platform. This is a valid test if you use an electronic warbler or no ringer at all on the test platform, but not a mechanical ringer, as there would be two mechanical ringers connected in parallel and the switch room would have a hard time driving both. This is another reason to stay away from a mechanical ringer on the test platform.

A.7 **Testing a hook switch.** The hook switch can be tested just using a volt-ohm meter connected to two terminals of the hook switch and see if it is open when the hook is "hung-up", and shorted when it is not "hung-up".

A.8 **Testing a network.** There is no test for a network other than to try it in a circuit where all the other components are known to work. Since they were made in the 60's and 70's, they are not very old and should be OK, but you never know. The WE model 101A networks are older but still should be OK, if they don't look like they have been abused. The main failure mechanism in the old coils, ringers, and magnetos is the permanent magnets lose their magnetism. This should not happen in the newer networks as the magnetic material is much better than that used in the old coils and it doesn't degrade nearly as fast.

A.9 **Testing a touch tone dial.** A touch tone dial may be tested using the platform shown in Figure A-2. Just hook it up according the instructions given in paragraph 2.13, and add terminals S and T as described in paragraph 2.8.

APPENDIX B.
REFURBISHING DIALS

B.1 **Two types of rotary dials.** There are two types of dials; WE dials, made by the Western Electric company, and AE dials, made by the Automatic Electric company and some other companies. For a more extensive write up on the difference between the dials refer to Appendix C.

B.2 **The connectors on each type of dial.** Figure B-1 shows the back sides of the WE and AE dials. The WE dial has 5 connections and the AE dial has 4. For hookup when following the wiring diagrams detailed in other chapters, use the connections shown in Figure B-1.

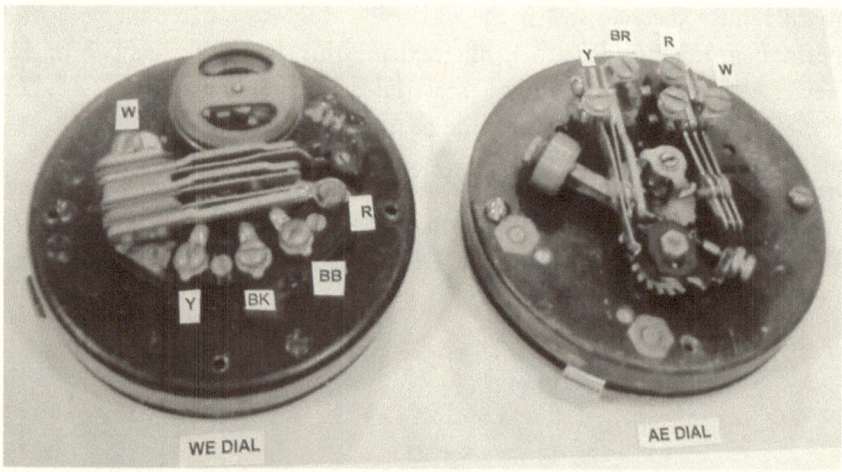

Figure B-1. Back-side view of the WE and AE dials showing the letter connections used to connect the dial to the telephone circuits.

Some AE dials have two connectors on the impulse side, Y and BR, and five on the shunt side as shown in Figure B-2. In this case you use only the two as shown in the figure labeled R and W. Be sure to check these out to make sure they are the right connectors by using your volt-ohm meter to see if they are open when the finger wheel is at rest and short when it is rotated from the rest position. If not, choose the correct connectors.

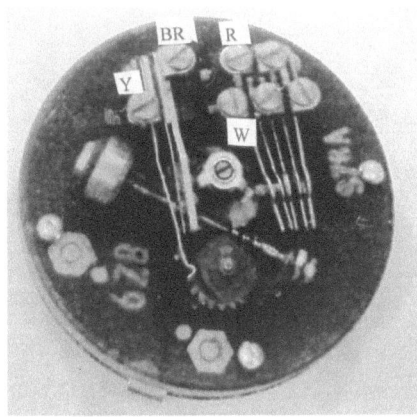

Figure B-2. An AE dial with five connectors on the shunt (right) side. Use only the two labeled R and W.

The impulse connectors are labeled Y and BR on the left.

B.3 **Labeling the connectors.** There are at least three different styles of WE dials, but you won't have a problem determining what the connections are as they are always factory labeled. The one shown above is the model 4H dial. For the AE dials, the connectors are never labeled so you need to determine which ones they are by looking at the switches. The impulse switch is the one that is operated by the rotating wheel which causes the switch to open and close several times as the finger wheel winds down. This is the switch that transmits the impulses to the switch room to tell it what the number is that the user is dialing. The two connections are labeled Y and BR (see the AE dial in Figure B-1 and the AE dial in Figure B-2). The shunt switches are the ones labeled R and W. These switches are used to disconnect the receiver from the rest of the telephone during dialing so the "popping" noises which are the dialing impulses will not be heard in the receiver.

B.4 **Removing the finger wheel on the AE dial.** Quite often the dials on old telephones are somewhat or totally stuck and will not wind down quickly enough. There is a certain range of wind down speeds that the switch rooms will recognize as dial pulses. If it is too slow, the dial will not work. You can usually correct this by taking the dials apart and cleaning and oiling the bushings. To get the finger wheel off, refer to Figure B-3.

Figure B-3. WE dial on the left and an AE dial on the right.

To take the finger wheel off you need to take the dial center off. For the WE dial, it is easy as you just put a small slotted screw driver on the top rim of the ring that holds the dial center on and pull it towards you to free the little clip that holds it. For the AE dial, insert a small slotted screw driver under the dial center rim at about the number 5 position and slide it down to about the number 7 position. This will remove a tab that holds the rim down. See Figure B-4, right picture. In the very center is the screw that holds a moveable slip ring now pointing to the left. With the small screw driver, you can slide the slip ring counter clockwise, from pointing to the #5 hole to pointing to the #7 hole. The slip ring will cease to hold down a tab on the dial center and you can pull the dial center off.

Figure B-4. WE (left) and AE (right) dials with dial centers removed and placed in front of the dials.

B.5 **Further disassembly of the WE dial.** The WE dial can be further stripped down by removing the nut and spring washer. Then remove the finger wheel. The AE dial has a slip ring that you moved when you unlocked the dial center ring using the screwdriver. Unscrew the screw, remove the slip ring, and remove the finger wheel. Do not attempt any further dismantling of either dial. With the finger wheel removed, you can spray paint the finger wheel and clean off decades of dust and grease from the number ring. If some of the numbers on the enameled number ring have been worn off through years of use, you can re-paint them with an artist's brush if you are a really nimble artist. Also you can get at both sides of the dial's working mechanism to lubricate it. To do this, spray the mechanism with WD 40 and rotate the dial back and forth many times, using a little force to get it to move faster. When it is working fairly freely, blow the WD 40 off with the canned air. Then put clock oil on all the bushings and speed control mechanism. Again, rotate the dial many times, sometimes over a 2 or 3 day period. You can buy a vial of clock oil from any clock repair shop. Do not use household or sewing machine oil as it will eventually dry out and become too viscous. Clock oil is especially designed to stay pliable for long

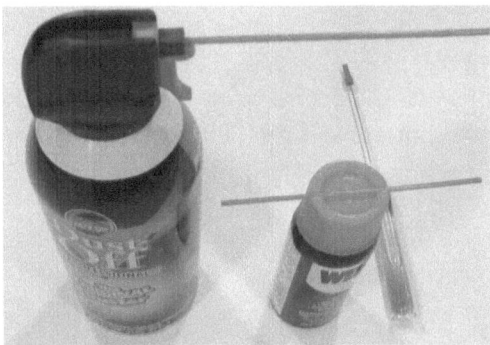

Figure B-5. (Left to right) Canned air for removing WD 40, WD 40 lubricant, vial of clock oil.

periods of time. Figure B-5 shows the items needed to clean and lubricate the dial.

B.6 **Putting the dials back together.** To put the dials back together, the WE dial is easy as you just reverse the process of taking one apart. The AE dial is a little different in that, when putting the dial center back on, make sure that the tab on the slip ring is in a position to fit *over* the tab on the dial center ring. Then, using the small slotted screw driver, move the slip ring from the number 7 position to the number 5 position, just the reverse of the process to take the dial center off. If you have trouble doing this, you might have to slightly bend either the tab on the slip ring or the tab on the dial center ring so the slip ring tab is on top. Since you have already seen how the parts work, you should be able to use your intuition to get the dial center ring to slip over the dial center tab and lock it back on.

B.7 **Dial repair specialist.** If, after the above cleaning and oiling procedures and the dial still does not work, you need to send it to a dial repair specialist. To adjust the timing on the dial takes some very specialized equipment that only a specialist would have. Do not throw it away as dials are a diminishing resource and a dial repair specialist can repair it for much less than the cost of buying another. Refer to Appendix J for the address of a dial repair specialist. The cost is very nominal, but you do have to pay for shipping and handling both ways.

B.8 **Bending the leaf springs.** Sometimes the switches on the back of the dial are bent a little and they need bending back to make sure they make and break contact with each other when needed. With this, you are on your own as it depends on which leaf springs need bending and by how much and in what direction. Use tweezers to grab the leaf springs and bend them. Watch them open and close to make sure they work as shown in Figure C-2, the timing diagram. The spring contact points may need cleaning. Use very fine emery cloth or sand paper, such as grit 220 or finer. Put it in between the two contacts and rub lightly. Then turn the cloth or paper over and repeat.

APPENDIX C.
THE FUNCTION OF DIALS IN
TELEPHONE CIRCUITS

C.1 **The two basic dials.** There are two basic dials used in old phones; the WE style dial and the AE style dial. The WE dial is made by the Western Electric company, The AE style dial is not only made by Automatic Electric but several other American and foreign manufacturers. The AE dials and others are all electrically the same but the WE dial is electrically different in that it;

- Connector R shorts out the transmitter when dialing so the dial pulses don't have to go through the transmitter element to get to the switch room. This is not done in the AE style dials and it makes little difference anyway.
- When dialing, every dial has to disconnect the receiver from the dial pulses or they create an annoying "popping" sound in the earpiece. Each type of dial does this in different ways but each gets the same result. The AE type dial shorts out the receiver element by shorting connectors W and R. The WE dial opens one line leading to the receiver so the pulses never get there. Both ways are equally effective. There is just no standardization.

C.2 **Action of the shunt switches for both dials.** Figure C-1 shows the circuit diagram for the receiver element and shunt switch for both the AE and WE dials

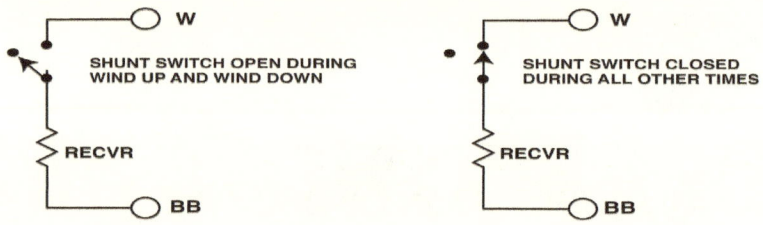

FIGURE C-1a. ACTION OF THE SHUNT SWITCH FOR THE WE DIAL

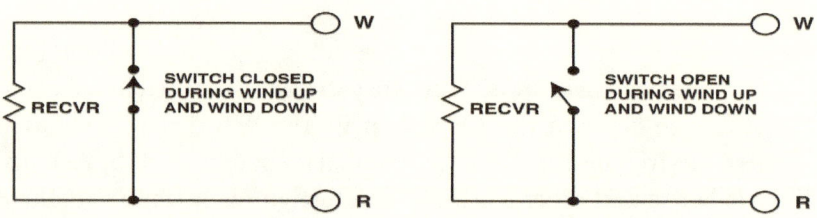

FIGURE C-1b. ACTION OF THE SHUNT SWITCH FOR THE AE DIAL

Figure C-1a. Action of the shunt switch for the WE dial.

Figure C-1b. Action of the shunt switch for the AE dial.

For the WE dial in Figure C-1a, the shunt switch is open during wind up and wind down, and closed during all other times. This action removes the receiver from the circuit during wind up and wind down preventing the dial impulses from going through the receiver. For the AE dial in Figure C-1b, the shunt switch is closed during wind up and wind down, and open during all other times. This similarly prevents the dial impulses from going through the receiver. Thus the ear is saved from having to listen to the popping sounds of the impulse switches.

C-3 **Timing diagram for dials.** Figure C-2 shows a timing diagram for both types of dials during wind-up and wind-down. There are four timing sequences drawn on the same set of axis so they can be compared; the impulse switch, the transmitter bypass switch, the AE shunt switch, and the WE shunt switch. Each is discussed below.

- IMPULSE SWITCH. The impulse switch stays closed during wind-up but opens and then instantaneously closes to send a series of high voltage impulses to the switch room which make up the number for that dialing position. The number of openings and closes indicates the number dialed.

- TRANSMITTER BYPASS SWITCH. This switch closes during wind-up and wind-down to short out the transmitter element and primary winding so the impulses generated above can go directly to the switch room.

- AE SHUNT SWITCH. During wind-up and wind-down, connectors R and W are shorted so the dialing impulses do not send an annoying "popping" sound to the receiver.

- WE SHUNT SWITCH. During wind-up and wind-down, connectors W and BB are opened so the dialing impulses do not send an annoying "popping" sound to the receiver.

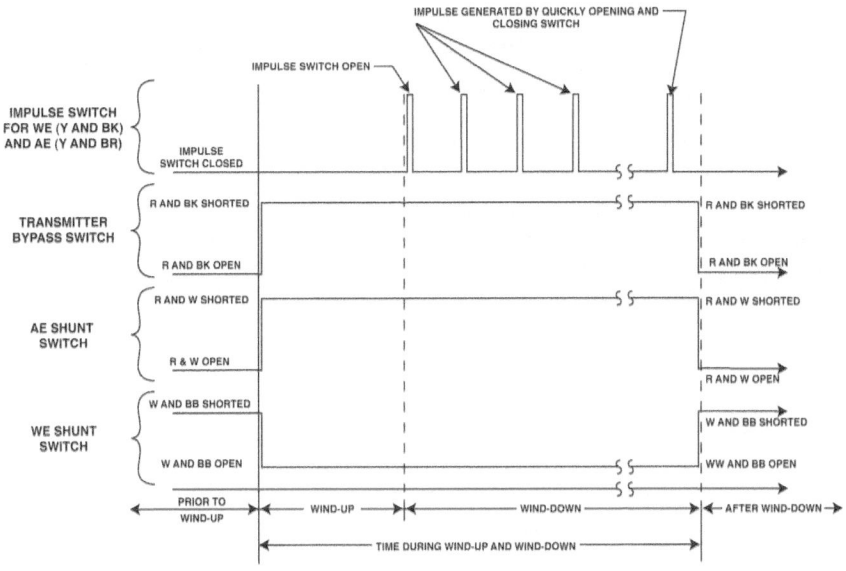

Figure C-2. Timing diagram for the impulse switch, the AE shunt switch, the WE shunt switch, and the transmitter bypass switch for a WE and AE dial.

C.4 **Comparison of the switch functions.** Figure C-3 shows a comparison of WE and AE dial switch nomenclature's purpose and function. In short, it tells what each connector is connected to in the circuit and what it does. This helps to clarify the purpose and function of each connector on an AE or WE dials.

DIAL		PURPOSE	FUNCTION
WE	AE		
Y	Y	ONE SIDE OF IMPULSE SWITCH	SPIKES OPEN AND THEN CLOSED FOR EACH PULSE DURING WIND-DOWN CLOSED ALL OTHER TIMES
BK	BR	OTHER SIDE OF IMPULSE SWITCH	
W	W	ONE SIDE OF SHUNT SWITCH	THE WE DIAL OPENS DURING WIND-UP AND WIND-DOWN (NOTE 2) SHORTS DURING ALL OTHER TIMES
BB	R	OTHER SIDE OF SHUNT SWITCH	THE AE DIAL SHORTS DURING WIND-UP AND WIND-DOWN (NOTE 2) OPENS DURING ALL OTHER TIMES
R	(NOTE 1)	SHORTS OUT PRIMARY WINDING AND TRANSMITTER WHEN DIALING	ALLOWS DIAL IMPULSES TO BYPASS TRANSMITTER AND GO DIRECTLY TO SWITCH ROOM

NOTE 1. THE AE DIAL HAS NO CORRESPONDING CONNECTION TO THE WE "R" CONNECTION

NOTE 2. TO ACCOMMODATE THIS REVERSAL OF FUNCTION, THE SHUNT SWITCH IS WIRED IN SERIES WITH THE RECEIVER IN THE WE DAIL, AND IS WIRED IN PARALLEL WITH THE RECEIVER IN THE AE DIAL.

Figure C-3. Comparison of WE and AE dial nomenclature, purpose, and function.

APPENDIX D.
REFURBISHING MAGNETOS AND RINGERS.

D.1 **Description of a magneto.** Magnetos are the generators that, when cranked in the old days, sent the electrical signal to the operator at the telephone company switchboard to alert the operator that the customer wanted to make a call. Figure D-1 shows two magnetos, one with three bar magnets, and one with five. Every generator has a stator and a rotor: the stator being the 3 or 5 magnets which you can see as large metal U shaped bars in Figure D-1, and the rotor being magnets with windings around them that are located inside the stator bars. In a generator, which is another name for magneto, the rotor rotates near the stator, and this generates an AC signal causing the ringer to ring.

Figure D-1. Two magnetos, a 5 magnet one on the left and a 3 magnet one on the right. The crank is on the right of each and the electrical connections for the magneto switch are on the left side.

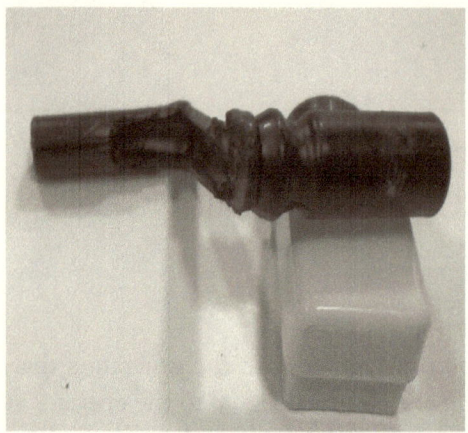

Figure D-2. Magneto crank. On the threaded side of the crank (left side), there is a split in the shaft (hard to see in this photo). For the larger shaft diameter, use a slotted screwdriver to spread the shaft apart. For the smaller diameter shaft, the crank will screw on with the gap closed.

D.2 **Oiling the bushings.** If you have an old phone, quite often the crank is missing or damaged beyond repair. If that's the case, new ones that look just like the old ones are available by mail order from parts suppliers listed in Appendix J. Fortunately, there are only two sizes of crank shaft diameters for American made phones, so the crank you get from the supplier will either screw on directly, or you might have to take a slot head screw driver and spread the threaded end apart so it fits the larger diameter crank shaft. Fortunately, the pitch of the threads on each is the same. See Figure D-2. Before turning the crank, it's a good idea to apply some lubricating oil to the bushings. Most magnetos have a funnel leading to the bushing, so just squirt the oil down the funnel. If there is no funnel, get some oil on the bushing as best as you can, and start turning the crank. Use 3-IN-1 oil, or any similar light household oil. Check to see that there is no debris inside the magneto that would hinder the rotor turning. If the magneto continues to be bumpy while you are turning after oiling, continue oiling it and turning it, sometimes for about an hour or two, taking rest periods in-between. Eventually, the new oil will displace the hardened oil and it will turn smoothly.

D.3 **Testing the magneto.** After you have the magneto oiled, connect a meter to the outputs and turn the crank to see how much voltage you get. It should be 70 to 90 AC volts for a fairly fast crank speed. If it's less than that, like about 35 to 45 volts, which is highly likely in the 100 year old ones, it will not have enough power to drive the ringer, and there is no fixing the magneto. The voltage output depends on how strong the magnetic field is in the bar magnets which compose the stator. The magnetic field was sufficient to generate 70 to 90 volts 100 years ago, but the magnets slowly lose their magnetism with age and there is no easy way to restore it. You'll have to connect the magneto to a modern day warbler which will work on a much lower voltage. Instructions for this are given in paragraph 8.8.

D.4 **Testing the ringer.** Next, you must see if the original ringer still works. Two things can go wrong with the ringers. Either there is a break in the wires that go around the magnets, which can easily happen as the wires are so small, or the permanent magnet in the ringer has lost its magnetic field, as in the case of the magnetos. Figure D-3 shows a mounted ringer.

Figure D-3. A mounted ringer on a wall mounted phone. You can see two solder tabs on the bottom end of the coils. Make sure the wires to the tabs are not broken by inspecting them with a magnifying glass or measuring the impedance between the two terminals with a volt-ohm meter. The impedance should be a near-short.

To see if the ringer still works, connect it to a telephone line using a test platform as described in Appendix A. If the ringer works, and the magneto works, you can connect them up as shown in Figure D-4 if you have a 3 terminal magneto. If either do not work, you must install a separate ringer for the rings from the outside line, and a warbler connected to the magneto to ring when the magneto is cranked. Refer to Chapter 8 paragraph 8.8 and Figure 8-9 for instructions regarding how to install and connect a warbler.

D.5 **The magneto switch.** The magneto switch is located on the magneto on the opposite side from the crank. There are two types of magnetos: 2 terminal and 3 terminal. Refer to Figure 2-11 and Figure 2-12 for instructions in wiring up either type of magneto.

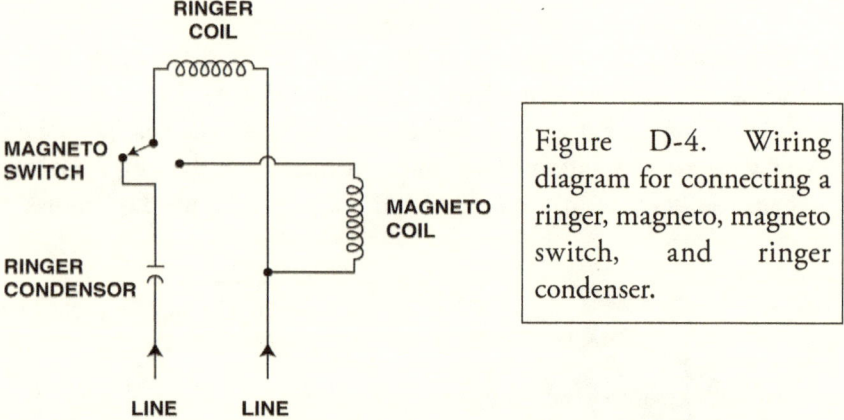

RINGER
COIL

MAGNETO
SWITCH

MAGNETO
COIL

RINGER
CONDENSOR

LINE   LINE

Figure D-4. Wiring diagram for connecting a ringer, magneto, magneto switch, and ringer condenser.

APPENDIX E.
REFURBISHING TRANSMITTERS AND
RECEIVERS

E.1 **Transmitter elements.** The old transmitters were made of carbon granules that were loosely packed when new, but with the passage of time the granules would stick together and become a hardened bloc. When they are hard, they don't work as well and the output is low, making it hard to understand the person using the phone. People want the clarity of modern day phones when using the old phones, so every transmitter element from a really old phone should be replaced with one out of a 1960's Western Electric model 500 or similar phone. You can get these out of the old phones or you can order them from any old telephone supply house for a nominal charge. The model T-1 or model N-1 are the transmitter elements you should use. The N-1 is smaller and can be used in handsets where there is limited space.

E.2 **Connecting to transmitter elements.** The transmitter element has always been the hardest telephone part for telephone manufacturers to deal with. Even with the newer elements, they cannot take the heat of soldering directly to them as the heat will cause hardening of the carbon granules. They might work OK for a while but they will eventually cause noise on the line which makes them intolerable to use. To overcome this, you can either buy from the suppliers in Appendix J two rings that will pressure fit around the two connectors, or make them yourself. Figure E-1 shows a pair of these connectors. The ones you buy already made up like in Figure E-1 are designed for the

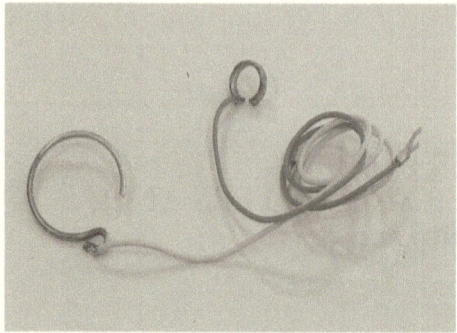

Figure E-1. Connector rings for connecting the two rings of a transmitter element by pressure fit to lead wires so you do not have to solder to them.

model T-1 transmitter element but they do not work on the smaller model N-1 elements. You can make them yourself for either size element by buying copper ferrules from any hardware store in the plumbing department. See Figures E-2 and E-3. Use a 3/8th inch size for the T-1 element and a 1/4 inch size for the N-1 transmitter. For the outer ring, use #18 or #20 gauge galvanized steel wire cut to a length just short of the circumference of the outer ring with a lead wire soldered to one end, and bent so it makes a pressure fit to the ring. The inner ring is the ferrule that has been cut along one side only and had a lead wire soldered to one end. This will provide a secure connection to the transmitter rings and will not heat them up while making the solder connections. Always make sure the inner and outer rings are not shorted together by inspecting for accidental shorts.

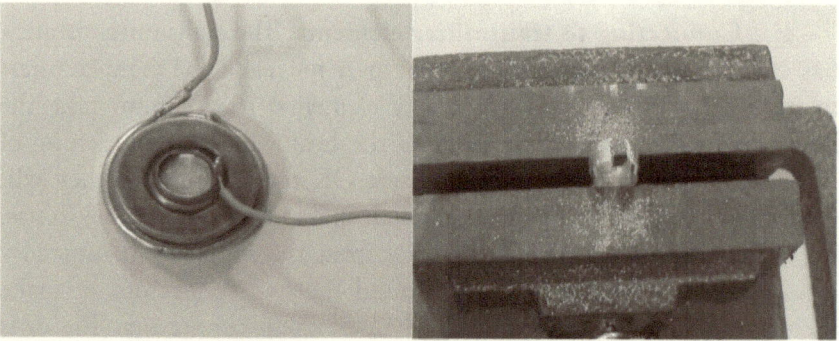

Figure E-2 (Left). Model N-1 transmitter with a ferrule around the inner ring and a piece of #18 gauge galvanized steel wire for the outer ring.

Figure E-3 (Right). Ferrule that has been cut with a hack saw on one side so it can have a lead wire soldered to it and then be pressure fit around the inner ring.

E.3 **Connecting the wired up transmitter element to the phone.**

The original transmitters never work and must be removed prior to installing the new transmitter element. This can be easily done by removing the transmitter face plate and removing the screws holding the old transmitter element.

The wired up transmitter element is now ready to be mounted on the phone. For an oak wall mounted phone, it is mounted at the end of the neck protruding from the front of the phone. Figure E-4 shows one ready to have the face plate screwed on. Note that paper towels are folded up behind it to give it a snug fit so it won't rattle. The wires go through a hole in the back of the paper towels and down through the neck and into the phone.

Figures E-4 and E-5. Figure E-4 shows the wired-up transmitter element mounted in the phone neck with folded up paper towels behind it to prevent rattling and insure a snug fit. Figure E-5 shown the face plate mounted on top of the transmitter.

The model T-1 transmitter element will easily fit into every American made phone but it is too big to fit into many European made phones such as most of the Ericsson phones. For these you use the model N-1 element as it is nice and small.

E.4 **Receiver elements.** A receiver element from 100 years ago may or may not work but it is certainly worth a try to salvage it. Try it and if the volume is low, try bending the metal plate as described in Appendix H. If not, the procedure to replace it is given next.

E.5 **Connecting to old receiver elements.** Figure E-6 shows the insides of an old oak wall mount telephone. The receiver housing is on the left. The middle part is the permanent magnet, the long tubular part, and the receiver mechanism attached to it. If it can no longer be made to work using the techniques of Appendix H, it must be replaced. The receiver cap is on the right.

Figure E-6. Insides of an old oak wall mount phone. The center element is the permanent magnet and receiver and must be replaced.

The old permanent magnet is what gave the receiver the weight to make it pull the hook switch down. So the weight must be replaced by using a 3 5/8 inch piece of ½ inch steel pipe, as shown in Figure E-7. The ½ inch steel pipe can be purchased at any hardware store and the pipe can be easily cut with a hack saw to 3 5/8 inches in length.

Figure E-7. Oak wall phone receiver with replacement parts consisting of steel pipe for weight and a modern receiving element

E.6 **Connecting to new receiver elements.** Most modern receiver elements have screw-down points to connect the wires to the receiver element. Put the wire through the receiver housing and through the steel pipe and connect to the receiver element. Use a cut up paper towel to make all the parts snug so they don't rattle and to insulate the wires from the pipe.

E.7 **Securing the earpiece cable.** When wiring the receiver element to the cable that will connect the earpiece to the oak wall housing, some way must be made to keep the cable from pulling out of the earpiece if it is dropped. The easiest way to do this is to just tie a knot in the cable so it rests just inside the hole of the housing. Another way is to wrap the wire with about five layers of #18 gage solid steel wire and press it on tightly with pliers. Either way, the cable will not pull out of the receiver, even if the receiver is dropped and the line cord stops the fall.

APPENDIX F.
REFINISHING THE WOODEN CASES OF OLD TELEPHONES

F.1 **Method used.** You can refinish the wooden case of an old telephone any way you like, but this is tried and easy guide in case you have never done it before.

F.2 **Removing parts.** Figure F-1 shows an oak ringer box that will be used in this guide. It doesn't matter if it's the entire oak phone or just the ringer box as the procedure is just the same.

Figure F-1. Oak ringer box and WE mode 202 telephone. The oak ringer box will be refinished as a demonstrator but the procedure is the same even if it is the entire phone such as an oak wall phone.

Disconnect the desk set from the ringer box and remove the bells and any other non wood parts. The small items such as the escutcheons, hinges, telephone line connectors, lightening arrestors, and hasps may be removed or just left on, as they are small and sometimes are difficult to remove and put back on. On this one, only the bells and the cable connecting the desk set to the ringer box were removed.

Figure F-2. Oak ringer box ready for refinishing. Note the blue masking tape over the original "Western Electric" logo so it will be preserved.

What is left is shown in Figure F-2. Note the old label saying "Western Electric" is taped over with blue masking tape so it can be preserved. It is possible to buy reproduction decals of these labels and glue them on after the refurbishing is complete but people generally like the original of everything if possible.

F.3 **Removing the old sealer.** The ringer box is now ready for stripping with paint stripper. I prefer the ones that are water soluble such as Stripeeze or Bix. You generally follow the directions on the can, which is:

- Brush the solution on getting it as thick as possible. Be sure to do it in the shade as it tends to dry quickly and if it dries it stops working. Be careful with the masking tape. Don't brush hard around it as will easily come off.

- After letting it rest about 10 minutes, brush some more on, again letting it rest about 10 minutes. Let the chemicals do the work, but don't let it dry out.

- Using a scraper, scrape off as much as you can. Again, protect the masking tape. Then brush on some more and let it rest about 10 minutes.

Now, all the chemicals should be ready for removal. To do this, use a solution of about one gallon of water with about 1/2 of a cup of TSP dissolved in it and a wad of coarse steel wool. TSP is a cleaning solution and in this case makes the paint remover dissolve in water much more readily. It is available in any hardware store or grocery store. The solution will do a good job of dissolving the paint remover if you rub it briskly with the steel wool. To get the last part off, take it out to the lawn and spray it with the garden hose with a strong spray, then wipe it with a cloth and let it dry in the sun. Again, be careful with the masking tape and the label underneath.

F.4 **Sanding.** When dry, sand it with 60 grit sandpaper to make it smooth. Then stain it with golden oak color wood stain and let it sit over night or longer. For a sealer, it is suggested that you use water based clear gloss or semi-gloss polyurethane sealer. Which one depends on whether you like a high gloss or moderate gloss finish. Stir this and brush it on and let it dry over night. Then sand it with very fine grit sand paper, about 220 grit, and apply another coat. This should give it a very refined finish as shown in Figure F-3. Note the original WESTERN ELECTRIC sticker has been preserved. It makes a beautiful furniture accent piece. Note the beautiful grain in the oak. This only appears with antique oak as it is from first growth oak, not re-harvested second growth oak that you find in new oak furniture.

Figure F-3. Refurbished ringer box with two coats of water-base clear gloss polyurethane sealer.

F.5 **Mounting the line cable.** The flat, four-wire cable going from the phone to the telephone wall outlet comes from the ringer box or from the back of oak wall mounted phones. This can be seen in Figure F-3 at the bottom center of the ringer box. A groove has been cut with a router for the cable to fit into so the box does not wobble on the wall due to the cable being in the way. This is a nice feature but not an absolute necessity.

APPENDIX G.
PAINTING THE NON-WOOD PARTS

G.1 **Parts to be painted.** All the non-wood parts except the cables should be removed from the phone and cleaned and spray painted a high gloss black. This includes the hand sets and phone bodies. Figure G-1 shows several parts that have been removed from the phone, cleaned,

Figure G-1. All non-wood parts except the cables should be removed from the phone, cleaned, and painted a high gloss black.

and painted black. Small escutcheons, lightening arresters, and tie down posts are hard to spray paint so those can be left on the phone unpainted and painted separately using a small artists brush and a separate can of high gloss paint intended for brushing on. You cannot spray some spray paint onto a newspaper and then use this to brush on, as the spray paint is much too thin. It will not cover correctly. It must be paint designed for brush painting.

G.2 **Procedure for spray painting.** Rub the non wood parts with a wad of fine steel wool or a cloth that has been dipped in a paint de-glosser or liquid sand paper that you can buy from any hardware store. When thoroughly dry, spray paint them with a high gloss spray paint. The hard part is to not let the paint run. Spray paint is very liquid, that is, very low viscosity, as it must be or the paint would not squirt out of the can. This causes it to run if too much paint is applied. Read the directions on the paint can, but generally you should put on several very thin coats instead of one thick coat. This is fairly easy to do as a thin coat will dry in a matter of 30 seconds, so you can put on another coat quickly, and keep doing this until you get a fairly thick coat. Then let it dry for at least 24 hours until the paint becomes hard. Note that the directions will probably say that you can put on second coats either within one hour or after 24 hours. My experience has been that even waiting 24 hours or much longer to re-paint, you may still get bubbles and ruffles from the second coat being over the first coat. So try to get it completely done with only one coat.

G.3 **The hook switch plungers.** When painting the body of a desk phone, be sure you don't get any paint down in the holes where the hook switch plungers go. The holes are just barely big enough to let the plungers move up and down and the paint will add some thickness causing the plungers to stick. Wad up some newspaper and stick it down the holes to protect them from receiving any paint. Likewise, do not paint the plungers.

G.4 **Painting the handset, the AE finger wheel, and the bells.** The handset is difficult to paint as there is no bottom to set it on while it is drying. So leave the cord connected and mask off a few inches of the cord so you can spray paint the rest of the handset and tie the cord to something to let it hang to dry. This is shown in Figure G-3, on the bottom of the figure. The center portion of the finger wheel on the AE dial should be masked off so that no paint gets into the center portion, as shown in Figure G-2. This is where the slip washer slips when putting on the dial center and the paint is thick enough that the slipping motion will be impeded. The WE finger wheel can be fully painted as it does not have a slip washer.

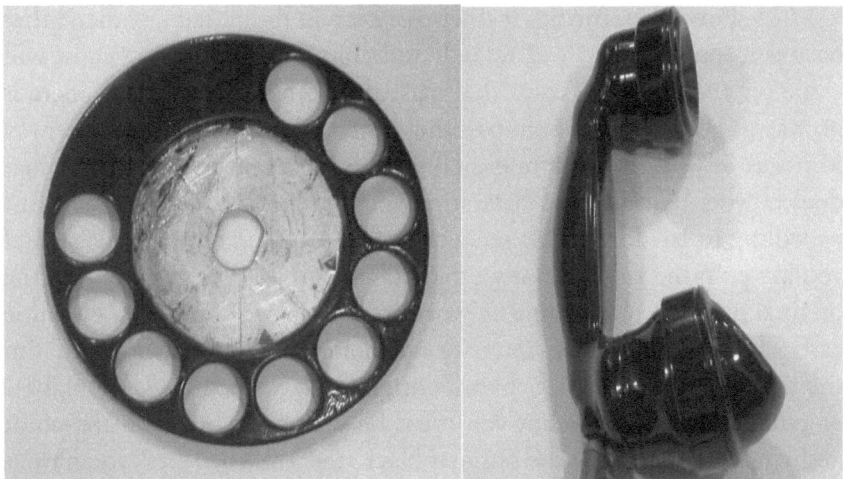

Figure G-2. (Left) The AE finger wheel. The center part has been masked off before painting so the thickness of the paint will not impede the motion of the slip washer. This applies to the AE dial only.

Figure G-3. (Right) Use masked off cable to the handset to hold on to it while painting and letting it dry. (The masking tape is at the bottom of the figure).

The bells were always made of brass and were always mostly painted black. Some restored phones have brass bells as some people today think brass is more beautiful. But in the original days when these were made, black was considered beautiful, not brass. So most restorers paint them black as they were originally. Sometimes, they made bells of brass with a nickel coating. These should not be painted black, just polished. Antique nickel develops a beautiful patina that you cannot get any other way and it should not be covered. Never have them chrome plated. Chrome was not commonly used in those days and it does not develop a patina with age. In my opinion, it is just not compatible with anything antique.

G.5 **Powder painting.** It is wise to keep in mind that any metal that has been spray painted will scratch and chip fairly easily. The paint will come off and show the color that existed before it was painted. There is no way to get around it, even painting it first with a primer. The only way to insure it will not scratch and chip off is to take it to a company that does powder painting and have it powder painted. Of course, you must be willing to pay for it. This scratching and chipping is only noticeable if you are painting the parts any other color than their original black. The original black on black painting works very well and even if you get some early chips and scratches, they are not noticeable. If you want to paint the phone any color other than its original black such as white or blue or pink, it can be done but you must be willing to live with the chips and scratches that show the original black underneath, unless you have it powder painted.

APPENDIX H.
REPLACING THE RECEIVER ELEMENT IN A WE MODEL E-1 HANDSET, AND BENDING THE DIAPHRAGM ON THE OLD MAGNETIC RECEIVER ELEMENT.

H.1 Replacing the receiver element in a WE model E-1 handset. The receiver housing on a WE model E-1 handset is very small and to replace it if the volume is too low, which is likely, needs special instructions. Always try bending the diaphragm first to see if you can make it work by following instruction in paragraph H.2, but if you can't, use these procedures to replace the whole element with a new one. Figure H-1 shows a model E-1 handset. Figure H-2 shows the old receiver element that needs replacing.

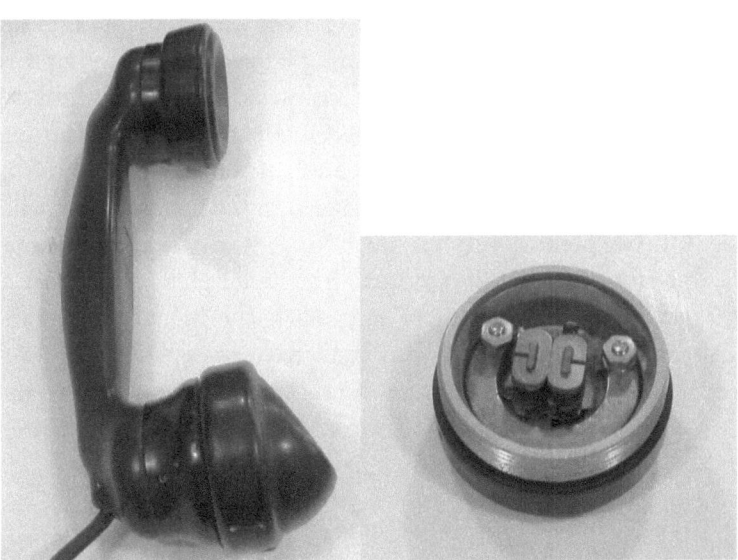

Figure H-1 (left). WE model E-1 handset.

Figure H-2 (right). Earpiece with cap removed and insides removed.

The housing is very small and will not take any receiver element except the smallest one which is the Audiosears model 2463V that you can buy from the supplier listed in paragraph J.10 in Appendix J. The insides of the receiver can be unscrewed and is shown in Figure H-2. Unscrew the two nuts shown in Figure H-2, lift out the contents, and grind down the shoulders of the housing as shown in Figure H-3. To do this, use a rasp as shown in Figure 8-8.

Figure H-3. Model E-1 handset with shoulders of the casing filed down with a rasp.

H.2 **Bending the metal diaphragm.** Before replacing any old receiver that uses the old electro magnet which vibrates a metal diaphragm such as the one shown in Figure H-4, try adjusting it to see if you can improve the volume. These old receivers do sometimes work very well and it is worth trying to save them if possible. To do this, unscrew the receiver cap, remove the metal diaphragm, and bend the center part of the diaphragm slightly towards the magnet that it sits next to when in the handset. Just a little is sufficient. If you bend it with too much force, it will develop a permanent crease, and if it does that, it is ruined. The metal plate should just touch the magnet. It you can get it to just touch the magnet, a remarkable improvement may result and you won't have to replace it. If you can succeed at this, remember that the metal diaphragm now has an up side and a down side and you must be careful when replacing it to make sure the correct side goes next to the magnets.

Figure H-4. Left, receiver element with magnets; middle, receiver diaphragm; right, receiver cap.

APPENDIX J.
LIST OF PARTS AND SERVICE SUPPLIERS

J.1 The top two suppliers (J.2 and J.3) can probably fulfill the majority of your parts needs so try them first. The third (J.4) has crimp-on connectors that fit old telephone wires such at the ones that connect the desk set to the hand set. Specify the model A29921CT-ND 18-20 AWG tin crimp. The VTS company (J.9) can repair rotary dials that are damaged beyond the scope of this book to repair and the Floyd Bell company (J.8) makes the warblers that work on telephones. For a warbler that works with telephones, specify the warbler model BR-3-39. Audiosears (J.10) makes the model 2463V receiver element that is only ½ inch thick required in the small European handsets. Phoneco (J.3) carries the model N-1 transmitter element which is small enough to fit in Ericsson handsets.

J.2 CHICAGO OLD TELEPHONE. MERGED WITH :
OLD PHONE WORKS AND HOUSE OF TELEPHONES
10 BURLINGTON COURT
KINGSTON, ONTARIO. CANADA
1 800 843-1320
WWW.OLDPHONEWORKS.COM

J.3 PHONECO
19813 EAST MILL ROAD
P. O. BOX 70
GALESVILLE, WI. 54630
608 582-4124
WWW.PHONECOINC.COM

J.4 DIGI-KEY CORPORATION
701 BROOKS AVE S.
THIEF RIVER FALLS, MN. 56701-0677
1-800-858-3616
WWW.DIGIKEY.COM

J.8 FLOYD BELL, INC.
720 DEARBORN PARK LANE
COLUMBUS, OH. 43085.
(614) 291 0823 (For telephones, specify model BR-3-39).
WWW.FLOYDBELL.COM

J.9 VTS INDUSTRIAL COMPANY
P. O. BOX 429
SALOME, AZ. 85348
520 370-3267.
JYDSK@TDS.NET

J.10 AUDIOSEARS CORPORATION
2 SOUTH STREET
STAMFORD, NY. 12167
800 533-7863 (For telephones, specify model 2463V receiver element).
WWW.AUDIOSEARS.COM

APPENDIX K.
REMOVING THE FINGERWHEEL ON THE
WE MODEL 500 TELEPHONE

K.1 WE model 500 phone. The telephone used in this chapter is a model 500 WE phone so it will be explained here how to remove the finger wheel on that model phone. It can sometimes be frustrating and tricky although it is mechanically straight forward, so lots of pictures are included here in place of a straight verbal explanation. To remove the finger wheel on a WE dial that is older than the model 500, refer to Appendix B, paragraph B.4.

K.2 Removing the finger wheel. Figure K-1 shows a model 500 phone with a tool that can be made from a paper clip to remove the finger wheel. Note the paper clip on the left hand side that has been partially straightened out leaving a handle on one end.

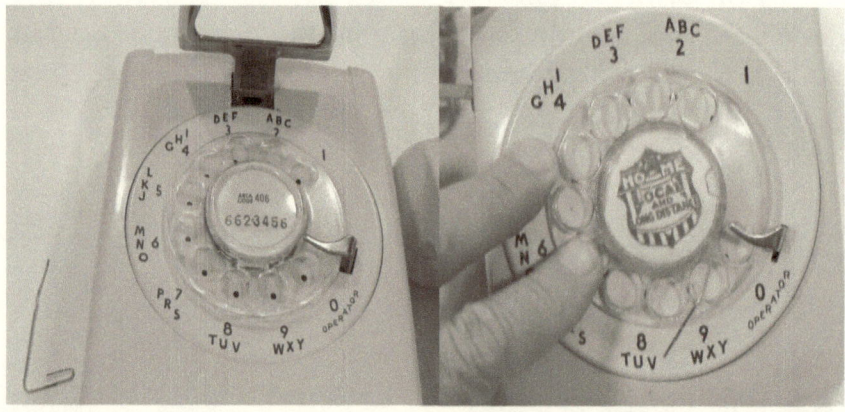

Figure K-1. Model 500 dial with tool made from a paper clip (lower left corner).

Figure K-2.Finger wheel stops with the 2nd finger hole in between the 3 and 2.

The first step is to rotate the dial clockwise (cw) until it stops. It will stop when the second finger hole is between the 2 and the 3 at the top of the dial. You will notice a small hole in the plastic finger wheel in between the numbers 9 and 0 as shown in Figure K-2. Figure K-3 shows a close-up with an arrow pointing to the small hole.

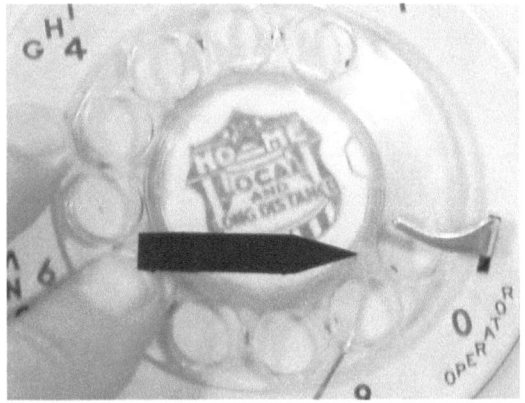

Figure K-3. Dial with black arrow showing the small hole where the paper clip tool is inserted.

It is right in between the second-to-last and last finger wheel holes.

Put the paper clip in the small hole, it is just big enough to accept a paper clip so it might be hard to find on your first try but it is there, and push the paper clip down hard while rotating the finger wheel in the cw direction. This will release a lever inside the finger wheel so it can be rotated to release it. See Figure K-4 to see the exposed lever that you are pushing on. It is a "wish bone" like spring with a push tab in the middle that the paper clip tool presses on. This releases the finger wheel so you can rotate it cw and disengage it from the axel.

K.3 **Putting the finger wheel back on.** To put the finger wheel back on, you don't use the paper clip tool. Simply position the finger wheel over the axel about in the position shown in Figure K-5 with the second finger hole over the 1 dot. Make sure it is flat, not hung up on one side. Then turn the finger wheel counter clockwise (ccw) until it snaps into place. You will feel a definite snap. When you are done, the first finger hole will be over the 1 dot as it should be.

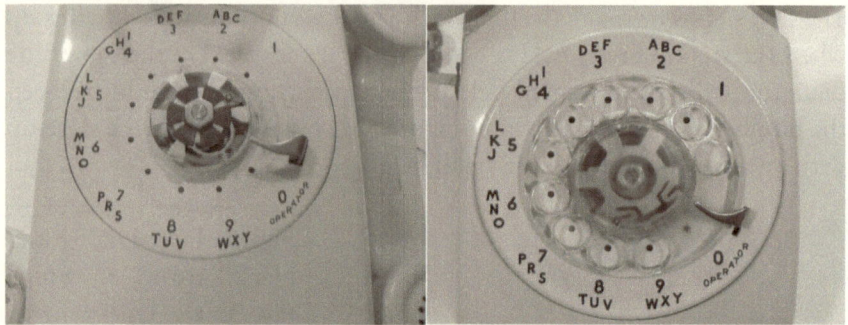

Figure K-4. Underneath portion of finger wheel showing the wishbone like spring with a push tab in the middle. The push tab points right to the middle of the 9 and 0.The push tab is what you press on using the paper clip tool, thus releasing the finger wheel so it can rotate cw to release the finger wheel.

Figure K-5. To replace the finger wheel, insert it on the axel at about the position shown in the figure with finger hole #2 right over the 1 dot. The finger wheel should insert all the way down so it is flat, not held up on one side and down on the other. Then rotate the finger wheel ccw until the first finger wheel is over the 1 dot. You will feel it and hear it click in to the lock position.

As you probably know from experience, some things don't always work out textbook style. If you are having difficulty pushing the wishbone tab down and you think the paper clip tool is not on the tab, drill out the hole in the finger wheel to make it just a little bit bigger so you can see through it and find out where the tab is. Then continue. If the drill hole is not too big, it will not be noticeable when the phone is finished.

APPENDIX L.
REFURBISHING TOUCH TONE DIALS

L.1 **Components of the dial.** Touch tone dials consist of coiled springs, magnetic coils, switches, resistors, diodes, capacitors, a transistor, and a few other components, all of which are quite reliable by themselves except the switches. The switches are out in the open and can get dust and dirt on them and the surfaces can oxidize over time. So to refurbish them you have to clean up the surfaces where the switches make and break the connections.

L.2 **Location of the switches.** Figures L-1. L-2 show the 12 switches in a model 35 touch tone dial that can go bad over a period of time.

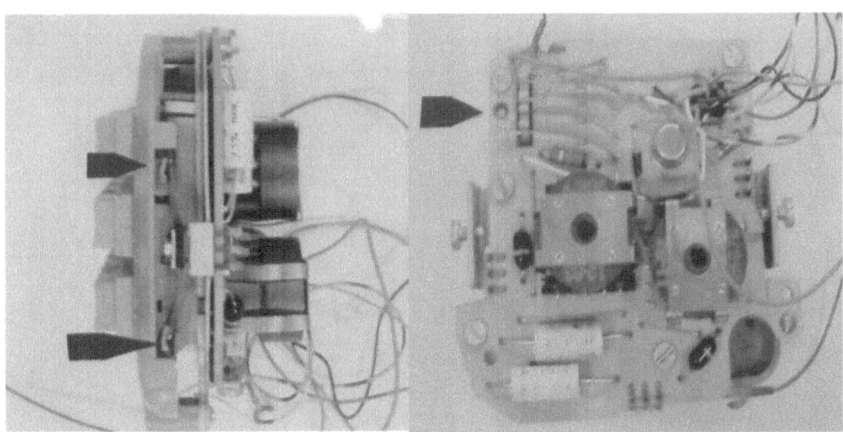

Figures L-1 and L-2. Figure L-1 shows the leaf springs on one of the edges of the dial of which there are four, and Figure L-2 shows the five leaf springs on the back of the dial as can be seen on the upper left quadrant. Black arrows show where the springs are in each figure.

L.3 **Cleaning the switches** To refurbish them, spray them with a contact cleaner such as DEOXIT which you can buy at Radio Shack or other electronic parts stores. The contacts are hard to clean and you have to exercise the switches multiple times after spraying them to loosen the oxidized layers. Repeat this several times. It might be necessary to do this the next day also, so the cleaner has time to work overnight. Keep repeating this process for all the switches until you have a working touch tone dial. If you have a model 72 touch tone pad, the switches are enclosed inside of the module and they cannot be cleaned, but the enclosure prevents them from becoming oxidized as fast.

L.4 **The transistor.** Probably the next most likely component to fail is the transistor. When testing the dial, if you can hear a continuous tone of any type when a key is being pushed, you know the transistor is good, as it, along with the coils, are the components that generate the tone. If the transistor is bad, there is no way to save the dial.

ABOUT THE AUTHOR

Mr. Mitchell has a Masters degree in electrical engineering and has worked in the aerospace field for fourteen years as an electronics engineer. He has collected and refurbished old telephones for ten years He is retired and lives in Orange County, California with his wife Susan.

REFERENCES AND ACKNOWLEDGEMENTS

REFERENCES   The following book is considered the standard for referencing technical material regarding old telephones. OLD TIME TELEPHONES; TECHNOLOGY, RESTORATION, AND REPAIR. By RALPH O. MEYER. TAB BOOKS Division of McGraw Hill, Inc. Blue Ridge Summit, Pa. 17294-0850.

ACKNOWLEDGEMENT   The author wishes to acknowledge Richard Marsh, former owner of <u>CHICAGO OLD TELEPHONES</u>, for his help in answering questions and giving guidance to the author while he was learning the techniques of refurbishing old telephones.

www.ingramcontent.com/pod-product-compliance
Lightning Source LLC
Chambersburg PA
CBHW022001170526
45157CB00003B/1096